TO THE MOON AND BACK

Essays on the Life and Times of Project Diana

Cindy Stodola Pomerleau

To the memory of my father, King Stodola, with the hope that this volume will help make the 75th anniversary of what he regarded as his life's greatest achievement a celebration worthy of its importance. It's been a long time coming.

And to my husband, Ovide Pomerleau, without whose help and support not only this book but the five years of blogging that spawned it would not have been possible. I am beyond grateful.

CONTENTS

INTRODUCTION

L t. Col. Jack DeWitt, head of the Evans Signal Laboratory at Camp Evans in Belmar, New Jersey, had long dreamed of bouncing radio waves off the moon. Indeed, he had already made a failed attempt to do so three years before his assignment to Camp Evans.

Following the Japanese surrender in 1945, DeWitt's bosses in Washington turned their attention to the next item on their worry list. Anticipating the development by the Soviet Union of long-range missiles capable of delivering nuclear weapons, the Army wanted to know whether such missiles could be detected by radar, even if they were to arc high into the atmosphere en route to their target. Accordingly, DeWitt was tasked with determining whether radio waves could pass through the charged layer of the earth's atmosphere known as the ionosphere, which extends from 50 to 600 miles above the earth's surface.

To understand why this might have been controversial, it is helpful to know that historically, long-range wireless communication depended on so-called "skywave" propagation, which makes use of the ionosphere as a reflecting surface for bouncing or skipping radio waves to distant locations. Indeed, transatlantic communication would not have been possible without skywave transmission because the curvature of the earth blocks line-of-sight transmission. Dewitt and others believed they could get past the ionosphere by using shorter and more powerful radio waves, but some of their contemporaries were skeptical.

No long-range missiles were available to settle this dispute...
but it might be possible to bounce a signal off a stand-in, a large
object outside the earth's atmosphere.

Say, the moon.

DeWitt quickly made his decision. He had the team, he had
the equipment, he had the expertise—and it was all about to be
disbanded, its collective capability lost forever. The offer of an
excuse to pursue his dream of "shooting the moon" was just too
good to pass up. In the waning days of their wartime service,
De Witt's small team—Chief Scientist E. King Stodola and three
more engineers, Herbert Kauffman, Jack Mofenson, and Harold
Webb, with input from other Camp Evans units including most
notably the mathematician Walter McAfee—began to work fe-
verishly on what DeWitt had code-named Project Diana, after
the Roman goddess of the moon, initiating the tradition of naming space missions after Greek and Roman deities.

On January 10, 1946, at 11:48 am, they aimed their purpose-
built antenna (cobbled together from two SCR-271 fixed radar
antennas mounted side-by-side on a 100-foot tower) at the
horizon and broadcast a series of radio signals. Their first few
tries were unsuccessful, but at 11:58 am, the moon began answering—tentatively at first, then definitively. The conversation
continued until 12:09 pm, when the moon moved out of radar
range. On the following three days, and on eight additional days
during the month, two-way communication resumed: Signals
were sent, and around 2.5 seconds later, the time it took to make
the approximately 480,000-mile round trip, the moon reflected
back the greeting from Planet Earth.

And the crowd went wild! News of their accomplishment
made banner headlines and flashed on movie screens across
the country and around the world. The local newspaper, understandably preening itself on being right there on the spot,
described it as "one of the most brilliant scientific accomplishments of all time." Anyone old enough to have lived through
the Apollo 11 era, and to remember how that "one small step
for man" captured the public imagination, probably has a fair
idea of the impact made by Project Diana. Parades were held to
honor this new breed of scientist-hero. Many were excited by

the glitz or stirred by patriotism. A smaller number, perhaps, grasped the full significance of the event—that this tiny handful of scientists had ushered in a new age, an age in which we were no longer bound in theory or in fact by the earth's atmosphere. The sky was no longer the limit. Project Diana demonstrated conclusively that the earth's atmosphere could be pierced and that communication with extraterrestrial bodies was possible. It marked the birth of radar astronomy and initiated the age of space exploration.

It also ushered in the Cold War.

* * *

Several years ago, I launched a website called *Project Diana: The Men Who Shot the Moon*, hoping to capture—in their own words and those of their colleagues, friends, and families—the human drama, the story of the men who made it possible and the chemistry among them that made it succeed. On January 10, 2016, the 70th anniversary of Project Diana's successful moonshot, I added a blog to the site titled (what else?) "To the Moon and Back." I originally conceived of it as a tribute to Col. DeWitt, to his chief scientist King Stodola, to the rest of their team, and to the many Camp Evans colleagues on whose additional contributions the team depended. I also hoped to attract some belated recognition for the feat. My father, King Stodola, had lobbied with only lukewarm success for 40th and later 50th-anniversary commemorations. I felt he deserved better and hoped to find a satisfactory explanation for the Army's oddly persistent reluctance to make a big deal of this stunning achievement. Spoiler alert: Eventually, with a little nudge in the right direction by one of my blog readers, I believe I actually succeeded in solving this mystery. (Check out my essay titled "The Moon Enters the Cold War").

But always, in the back of my mind, lurked the niggling sense that there was more to be said, and that Project Diana might constitute an interesting lens for examining some of the transformations and dislocations occurring in the U.S. in the aftermath of World War II. Project Diana and its culmination on that chilly

winter morning occurred at a singular nexus in American history—looking back on our emergence from isolationism and the
deprivations and horrors of World War II, and ahead to the Baby
Boom, suburbanization, the Cold War, radical changes in the
lives of both men and women, communications breakthroughs,
and an astounding acceleration in the pace in scientific development.

Since then, I have written dozens of essays not only on the
technical and political aspects of the moonshot but also on sub-
jects ranging from my father's character and his childhood family to my ukulele and my Toni doll. Everything of the period has
been grist for my mill. My only organizing principle has been to
vary my fare in hopes of providing something for everyone; if I
wrote an entry on antenna design, look next for one on Spam or
on how my parents met. Researching all these topics from my
own particular perspective, sometimes asking questions that
apparently no one else ever bothered to ask, has occasionally led
me to surprising conclusions. If anyone had told me when I first
made Project Diana my mission that I would be editing Edwin
Howard Armstrong's Wikipedia entry to include little-known
information on his World War II research, I'd have laughed in
disbelief.

January 10, 2021, marks the 75th anniversary of Project
Diana. The book you are reading represents my contribution to
the celebration of this milestone. Although many (not all) of
these essays started life as blog posts, I doubt that even the most
devoted followers of my blog will feel cheated. The essays have
been extensively revised and updated and in some instances
completely rethought. I have also arranged them by topic—and
within topic, roughly chronologically—editing for consistency
and adding new material to help connect the dots. I have tried to
minimize the repetition that became apparent when essays originally published months or even years apart were arranged by
topic, cheek by jowl; if I missed a few instances, I trust at my age
I can be forgiven for occasionally telling the same stories over
again.

* * *

I believe I'm reasonably well-poised to chronicle the life and times of Project Diana. Although I'm not a professional historian or sociologist, I grew up in Diana's shadow on the Jersey shore and have numerous documents, photographs, and clippings pertaining to my father's work. My earliest playmates were other Project Diana and Camp Evans "legacies." Unlike my father, who professed to recall little about his childhood, I remember my own vividly. And although I'm not an expert in radio, radar, or Earth-Moon-Earth communication, I boned up enough to earn my amateur radio technician's license and reclaim my father's call sign, W2AXO. I also know enough to call for help when I'm over my head, and it was generously given—much of it by my husband Ovide, a licensed radio amateur for nearly 70 years, but also by authorities in other areas, as detailed in the Acknowledgments at the end of this book.

PART I: PROJECT DIANA AND ITS PLACE IN HISTORY

A. WORLD WAR II

THE DAY EVERYTHING CHANGED

E arly on the morning of December 7, 1941, the Japanese launched a surprise attack on the U.S. naval base at Pearl Harbor in Hawaii, sinking or damaging 18 ships and destroying 188 aircraft. In all, 2,403 were killed—mostly service personnel but also including 68 civilians—and 1,178 wounded. Since the U.S. was not at war, all the victims were non-combatants. For Americans then and ever after, it was, as President Franklin Delano Roosevelt famously referred to it, "a day which will live in infamy."

Congress quickly heeded Roosevelt's call for a declaration of war. Four days later, unhappy about Japan's unilateral, unannounced initiation of hostilities but realizing American participation in World War II was now inevitable, Germany and Italy declared war on the U.S., which immediately reciprocated.

Although the U.S. had kept a wary eye on developments in Europe (Asia not so much), until now it had maintained a staunch neutrality, its populace deeply divided on whether America should be involved in the war effort in any way. All that changed with Pearl Harbor. Within hours America was on a wartime footing. Soon films about military life and lovers separated by war would crowd out *Citizen Kane* and *Dumbo* in the movie theaters. Soon "Boogie Woogie Bugle Boy" and "I'll be Home for Christmas" would dominate the airwaves, along with revivals of World War I hits like "Over There." Soon deprivation and shortages of building materials and consumer goods would become the norm. Soon admonitions like "Remember Pearl Harbor!" and "Loose lips sink ships" would become part of our daily conversation.

Much has been made of the parallels between Pearl Harbor and 9/11. Both were unexpected attacks on iconic homeland targets, both inflicted a shocking amount of damage, both resulted in thousands of casualties, and both brought a sudden reveille to those who thought the world outside our borders could simply be ignored. Both drew the U.S. into many years of armed combat. Whether the long-term ramifications of 9/11 can possibly match the political, economic, sociological, and cultural dislocations that followed World War II—the decades-long dominance of the U.S. on the world stage, the Cold War, the increasing pressure for gender and racial equality—remains to be seen. But the analogy is useful in giving those too young to remember Pearl Harbor at least a hint of its transformative effect on life in America.

For both the men and women of the Camp Evans community, the impact of Pearl Harbor was, if anything, magnified by the circumstances in which they found themselves—that is, in a new, *ad hoc* community, with nothing in the way of roots or shared traditions; and with all the men suddenly on high alert, aware that if the Axis powers knew what was going on at Camp Evans. it too could become a prime target.

✻ ✻ ✻

My father's story was perhaps typical. His first job after graduating from the Cooper Union in 1936 with a shiny new degree in electrical engineering was as a "student engineer" for Pan American Airways, where his chief duty was painting antenna poles. After a series of "starter" jobs with gradually increasing responsibilities, he eventually landed in Washington, DC, in a Civil Service position in the Signal Corps, in which he advanced from Junior Engineer at $2,000/year to Assistant Engineer at $2,600/year. As the new decade dawned, my father, finding his job too long on administrative duties and too short on research (in a word, he was bored), sought a more stimulating position, one that would draw more extensively on his hard-won electronics skills. He was just on the verge of accepting a job at the Bureau of Standards in Washington in early 1941 when the Signal Corps countered by offering a promotion to Associate Radio Engineer at their radar laboratories at Fort Monmouth in NJ. He jumped at the chance.

Much of what I know about this era of my father's life was gleaned from an oral history interview I conducted with him in 1979. He told me his initial assignment was at Fort Hancock, "an isolated peninsula up near New York City; but we were eventually transferred to the old Marconi Radio Building down at Belmar [which the Army had only recently acquired]." To minimize his commute and allow him to ride his bike to work sometimes (they only had one car), my parents moved at around the same time from Long Branch, their first New Jersey home, to Shark River Hills.

If my dad was looking for excitement, he almost certainly found more than he'd bargained for. My parents and their friends were of course keenly aware of the war in Europe—how could they not be, given my father's line of work?—but it was someone else's war, not theirs. Preoccupied with unpacking their boxes, adapting to their new life, thinking about having children (a question of when, not if), they were as oblivious as most other Americans were to the fact that war was about to lap up to their own shores, and even more surprisingly, at the hands not of Hitler but of the Japanese: "Pearl Harbor...was a tremendous shock to us. I guess if we really stopped to think of it, we would have realized that something like this was inevitable, because Hitler's intentions were very clear—dominate the world!—and he would form whatever alliances and whatever he needed to do it. But when the shoe dropped, as it were, it was a great shock. It was a Sunday morning, and we were madly telephoning—how can we get out and man those radar sets and do something about it?—a panic, pretty much of a panic."

Several workplace changes resulted almost immediately from the attack on Pearl Harbor. Security, already tight, was strengthened even further. My father's work acquired a laser-like focus on the military applications of radar: "I eventually wound up heading what was called the Special Developments section with about 15 or 20 people in it and we did some very interesting work in radar [including the Army's first moving target search radar].... I think some of it was very original." Another dramatic change was in their work schedules: "Overtime became the rule rather than the exception—in fact, we worked pretty much a six-day week."

* * *

When the Camp Evans wives were parachuted into the unfamiliar Jersey Shore culture, they found themselves quite isolated, their social circle largely defined by their husbands' work ties. After the sky fell on December 7, 1941 and their husbands started spending more and more time at work, they clung to each other all the more tightly for solace and companionship.

My mother's closest friend from our Shark River Hills days was Mary Jane Evers, an outgoing woman with a wry sense of humor. Mary Jane had a way with words and for many years wrote an amusing column in the *Asbury Park Press* called "We Took to the Hills" that went well beyond the demands of the genre (which tended to focus on who had tea with whom or presided at the ladies' auxiliary meeting). In 1996, when I asked her to contribute to a "collective memoir" of my mother, she responded with a charming, breezy essay, scrawled in longhand, portraying both the impact of Pearl Harbor and her evolving friendship with my mother.

Although she was still living in West Long Branch at the time of Pearl Harbor, Mary Jane was obviously already tuned into life in "the Hills" and the Evans Lab community: "You must remember we were all 'strangers in a strange land,' so to speak. Our husbands had been assembled from all over to nurse the infant radar labs and then the electronics labs. None of us had family nearby; shortly after we met, the attack on Pearl Harbor [occurred]; we each had 4 gallons of gas a week for the family car, and meat and sugar rationing."

Sometime in 1944, when my mom and Mary Jane each had a toddler daughter and Mary Jane had recently given birth to a second child, the Evers family moved from West Long Branch to Shark River Hills. By late fall of that year, Mary Jane was going through a very rough patch. Her infant daughter had recently died at the age of three months, and her husband Jim, who like my father had joined the Radar Division at Evans shortly before Pearl Harbor in 1941, was away on "travel duty." Impulsively, my mother phoned and invited her to join them for Thanksgiving dinner. Their friendship was cemented with that gesture.

Mary Jane was keenly aware of how unlikely, in many ways, their friendship was: "We were very different people and from different backgrounds, but I believe that being thrown together in need, we grew to respect each other's views and to treat our children a bit differently [than we might otherwise have done]. She was very loving (almost doting with you!); I saw children as little beings who would grow, hopefully, into individuals I would like. I have certainly achieved that!" (To be honest, I have no idea what those differences might have been—from my child's-eye point of view, the two women were a perfectly matched pair of moms. Politics? Whatever the differences were, clearly they were apparent to Mary Jane and presumably to my mother as well.)

Aside from their children, the other perennial conversation

topic was money, or more accurately, the lack of it. According to Mary Jane, the pay was not generous even by wartime standards:

"Civil Service personnel were not among those getting raises in Congress. The general public, from what I have learned, figured they had had money all thru the Depression when everyone else was broke, so they could just wait."

"Budget problems were ever on our minds," she went on to say. "Your [parents] would have some 'interesting' discussions when the bills came in. My daughter Barbara vividly remembers hearing Elsa [my mother] say, 'There's always too much month at the end of the money!'" The women raided their kids' piggy banks (we all had piggy banks, which were supposed to teach us to save our money), and as my mother commented to Mary Jane, "By the time I pay the children back, I'm broke again!" Elsa and Mary Jane bartered babysitting time by deliberately joining organizations with different meeting schedules: "She was in the AAUW and I was in the League of Women Voters; she was a member of the local Fire Auxiliary and I was a member of the Hospital Auxiliary."

At the end of this litany, Mary Jane worried she might have left me with the impression that the lives of the Evans wives were all about "money-grubbing. Of course, none of us was suffering from malnutrition or doing without the basics of food, clothing, and shelter. The point was that money was needed not only to provide the necessities of life but, in a world filled with bad news and uncertainty, to allow for the comfort of a few extras—a spot left over for simple parties, cheap beer and soda and birthday cakes." Making things come out right, making ends meet, making it possible to have birthday cakes as well as Spam —that was a job that fell to the women. It was part of their contribution to the war effort, and they took it seriously.

❈ ❈ ❈

How did the children of the Camp Evans community fare during Pearl Harbor and its aftermath? I was born just over a year after Pearl Harbor and have only fleeting memories—if indeed they are authentic memories at all—of the war years. Many of the facts of wartime life were part of the air we breathed. Yes, we ate Spam for dinner. Yes, we observed blackouts and dim-outs to avoid bringing unwanted attention from the German U-boats to American supply ships. Yes, we shared in the American love affair with the radio and tracked the terrifying narratives it brought into our homes. Yes, a chronic state of low-level depriv-ation was a part of our daily existence.

But in some important respects we were sheltered. The men —whether they were in the military or, like my father, civilians employed by the military—were doing work deemed critical to the war effort and therefore spent the war years on the home front. They may have left for work early in the morning and ar-rived home late at night, but at least they were there, not thou-sands of miles away like my father-in-law, who spent four years on the Italian front as an Army surgeon. Our uncles and cousins may have been in uniform overseas, but not our fathers.

And perhaps partly for that reason, our moms were stay-at-home moms, quite unlike my stepmother Rose, who re-sponded to the wartime labor shortage by going to work as a welder. Although to say the Camp Evans wives bore the brunt of child-rearing responsibilities would be an understatement— and without benefit of Dr. Benjamin Spock, whose revolution-ary book on childcare wasn't published until 1946—at least "Just you wait till your father gets home" was not an empty threat. (Not that my own mother said that, ever, but I certainly knew children whose mothers did.)

So in a way we Camp Evans kids got the jump on the Fifties. At a time when other families were struggling to adjust and cre-ate a new normal, we were already there. The Baby Boom was already in progress. Sally, the oldest Evers child, was born on September 11, 1942—39 weeks to the day after Pearl Harbor. My mother suffered a miscarriage not long before I was conceived; otherwise my parents too would have had a "Pearl Harbor baby."

Perhaps that's why, whatever arbitrary cut-points the demographers adopt, I've always known in my heart of hearts that I'm a "Boomer."

Our mothers also didn't have to be hounded out of their jobs and back to domesticity, they'd been there all along. I'm not sure to what extent, if any, the resurgence of feminism rooted in the wartime increase of women in the workplace ever touched my mother. Much later she took a few education courses in the hopes of translating her college English major into a marketable skill, but by that time her health was starting to fail and her re-tooling scheme never got off the ground. Her ambition for her three daughters was that we should marry "well" so that we too could have the privilege of staying home to care for our children. We all remember her saying, only half-jokingly, "It's just as easy to love a rich man as a poor man." Fortunately for us, in light of subsequent economic shifts that made the one-income family a luxury, as well as our own ambitions, we got quite a different message from our father, who presented his women colleagues as role models and urged us to take all the math and science we could cram into our schedules.

My parents and their friends were part of what Tom Brokaw called the "Greatest Generation"—men and women born between 1901 and 1924, who came of age during the Depression and World War II, who shared a common core of values including honor, service, love of country and family, and personal responsibility, and who more than rose to the occasion when duty called. For better and for worse (don't forget the corrosive racism that permeated American society then and throughout our history, the internment or more accurately imprisonment of American citizens of Japanese descent, or the abuse my pacifist uncle suffered as a Conscientious Objector during World War II), they shaped America as we know it today.

As Brokaw observed, it was sometimes difficult to coax their stories from them because of their conviction that they weren't doing anything special, just honoring their commitments and doing what they were supposed to do. In this context, I feel fortunate to have obtained, without really planning to do so, the two eyewitness accounts, my father's and Mary Jane's, on which the above narrative largely rests.

TORA, TORA, TORA?

I t's an open secret—and a quintessential Camp Evans nail-biter—that the surprise attack on Pearl Harbor wasn't exactly a complete surprise after all, at least not in the sequestered halls of Camp Evans.

When I once casually mentioned that Army radar had in fact successfully detected Japanese planes approaching Pearl Harbor, a shocked friend e-mailed back, "WAIT! You're saying your dad... knew planes were coming in to bomb Pearl Harbor???.... That is a breath-taking piece of information, Cindy."

Lest I leave any room for confusion: No, my father had no idea that planes were coming in to bomb Pearl Harbor on the morning of December 7, 1941, as he later stated unequivocally in my oral history interview with him. In fact, there is no reason why he should have known. He was still a newbie at Camp Evans, less than a year on the job. His task was to optimize the equipment then in use, the SCR-270 mobile radar unit, and work on the next step in the series, the SCR-271 fixed radar, planned for installation at Pearl Harbor but not yet built.

In the hours during and after the attack, the only ones likely to be privy to information about what had or hadn't been detected were the members of a select, top-secret group that had been working at Camp Evans for years to prevent surprise air attacks on critical vulnerable targets including not only Pearl Harbor but also the Marshall Islands, Midway, the Philippines, the Panama Canal, and others. Needless to say, these men held positions well above my father's pay grade.

* * *

What actually happened in Hawaii that fateful morning is a classic example of an advanced new technology getting out ahead of its end users. Privates Joseph L. Lockard, age 19, and George E. Elliott, in his early twenties, reported for duty at 4 am at the Opana Mobile Radar Station, located 230 feet above sea level on the northern tip of Oahu. Although they were supposed to work in three-man teams, only the two men were on duty —Lockard serving as Operator and Elliott as both Plotter and Motorman.

It was an unusually quiet morning, and Lockard took advantage of the lull to train Elliott in the use of the SCR-270B mobile radar unit. At 7 am, the end of their shift, Lockard began shutting down the unit, when suddenly the oscilloscope picked up an image on the 5-inch screen so surprising he first thought something was wrong—a blip so large it must have been at least 50 planes. As of 7:02 am, the blip appeared 132 miles from Oahu. Elliott suggested they report this reading to the Information Center at Fort Shafter, around 30 miles south of the Opana Station. Lockard hesitated at first, but after several minutes of conversation—during which the blip moved another 25 miles closer to Oahu—he gave Elliott the go-ahead.

At the Information Center, Lieutenant Kermit Tyler, Pursuit Officer and Assistant to the Controller, was on duty that day, and except for the Switchboard Operator, he was alone. The Switchboard Operator took down Elliott's message—then, realizing that Tyler was still in the building, turned the call over to him.

Tyler's job description was "to assist the Controller in ordering planes to intercept enemy planes...." This was his second time serving in that capacity, the first having taken place three days earlier; he had no training in radar. The Controller and

the Aircraft Identification Officer were out of the building having breakfast. Dismissing out of hand the possibility that the
blip could actually be incoming enemy aircraft, Tyler scoured
his mind for alternative explanations, then remembered that a
squadron of B17 bombers—"Flying Fortresses"—was expected
from the mainland that morning. With a sigh of relief he uttered
five words that haunted him for the rest of his life: "Well, don't
worry about it."

By now it was 7:20 am. The planes were 74 miles away.

The first bombs struck Pearl Harbor at 7:55 am, and only then
did the three men realize what it was they had seen on the radar
screen. Had the information been passed along, even with only a
little over half an hour's lead time, American aircraft might have
been dispersed and ammunition readied. Had the Navy been no-
tified, it might have used the information to help locate the Japanese aircraft carriers from which the invading planes took off.
Although it is unlikely the main thrust of the attack could have
been averted, a response, *any response at all*, might have demor-
alized the Japanese by undermining their supreme confidence
that they had achieved "tora," a surprise attack—a goal they saw
as crucial to their success.

In subsequent inquiries, Tyler was exonerated due to his lack
of training and experience. Lockard received the lion's share of
the credit and was awarded the Distinguished Service Medal in
1942. Elliott was given the Legion of Merit but declined because
he felt, with some justice, that he should not be given a lesser
medal than Lockard.

❊   ❊   ❊

Meanwhile, back at Camp Evans, thousands of miles to the
east, members of the team charged with preventing surprise attacks waited on tenterhooks when they learned of the attack on
Pearl Harbor, fearing their radar had failed. "If our radar had not
given warning because of breakdown, or just ineffectiveness,"

said First Lieutenant Harold Zahl, "surely part of the finger of blame would point at our group." He himself had designed and hand-made special tubes for the radar set; had one of them failed? Not until several days later did they receive a call from Washington reassuring them that human error and not equipment failure had been responsible.

Although no one knew exactly when or where an enemy strike might occur, the Navy had developed and emplaced ship-based radar units as early as 1940, and by early 1941 the Army team at Camp Evans had set up land-based radar systems in potential target areas around the world. In addition to the Opana station, four other radar units had been installed in Hawaii. This is no secret.

Moreover, the story of the radar signals received by Lockard and Elliott but misinterpreted by Tyler and then ignored altogether is hardly an obscure anecdote buried in tomes read only by military historians or on websites visited only by passionate World War II buffs. On the contrary, it has been retold countless times in popular books and movies over the years. Notable among them: a brief, readable account titled *A Day of Infamy* written by Walter Lord and published in 1957 (60th-anniversary edition issued in 2001), in which a riveting description of the events that unfolded in the hour before the attack appears; a Japanese-American full-length dramatization of the events of that day titled *Tora! Tora! Tora!*, which garnered an audience score of 81% on Rotten Tomatoes; and a 1981 *New York Times* bestseller titled *At Dawn We Slept*, the first volume of a massive trilogy by a history professor named Gordon W. Prange, who devoted not years but decades to the study of a single day in history, generating thousands of typewritten pages of text that after his death were valiantly edited by two assistants-coauthors. (Lockard and Elliott make their first appearance 500 pages into the book.) The story even appears in *wikipedia*. As I said, an open secret.

* * *

So why is it that so many of us still cling to the myth that the U.S. was totally unprepared for the attack on Pearl Harbor?

Perhaps because the Japanese version of the story—that the attack on Pearl Harbor came as a complete surprise to the Americans—is the one that captured the world's imagination. Mitsuo Fuchida, leader of the first wave of Japanese fighters, famously sent the message "Tora, tora, tora!" to his superiors waiting on the aircraft carrier *Akagi*—making the communication (intentionally) puzzling to the casual listener since the word *tora* means "tiger" in Japanese. But "tora" was also a radio codeword combining the two Japanese words *totsugeki* and *raigeki*, a phrase meaning "lightning attack"; to those in the know, "Tora, tora, tora" had nothing to do with big cats and everything to do with having delivered a bolt from the blue. And why shouldn't the Japanese have believed this? After all, from their point of view there was no indication that anyone had the slightest inkling that an attack was underway. No defense was mounted, no evasive action was taken, thereby allowing the Japanese to punch above their weight at Pearl Harbor.

Or perhaps it's because we collectively prefer the metaphor of the sleeping giant awakened to the less heroic conclusion that three undertrained, inexperienced men had been entrusted with a new technology, and that but for human error, the encounter might have taken a somewhat different turn.

What is lost in the myth-making process is perhaps a minor footnote to the overall arc of the Pearl Harbor narrative but an important chapter in the history of radar. It was the first wartime use of radar by the U.S. military, and, despite the series of mishaps that rendered it useless at Pearl Harbor, it was abundantly clear that this revolutionary new technology was poised to transform the way war was waged. Being neither a military historian nor a radar scientist, I will leave a fuller investigation and interpretation of these developments to someone better qualified than I.

THE AMERICAN FRONT

Suppose they gave a war and nobody knew?

In the minds of most Americans then and now, World War II was "Over There." The home front was regarded not as a war zone, but rather as a base of operations for providing supplies and support. America was where Rosie the Riveter kept wartime production going so that our boys could go overseas where they were so badly needed. America was where homemakers like my mom, along with her friends Mary Jane Evers and Ruth Mofenson and all the other Camp Evans wives, got by on a ration of 4 gallons of gas per week so the War effort in Europe could be fueled.

And yet just days after Pearl Harbor, Hitler ordered an attack on America. He had been anticipating the entry of the U.S. into the War, which would likely infuse American industrial might into the Allied effort, and believed that preventing the U.S. from supplying Britain with fuel and arms would be key to a Nazi victory. The Nazis had been fighting what Winston Churchill dubbed the "Battle of the Atlantic" since 1939, including the use of *Unterseeboote* (U-boats) to attack Allied merchant shipping in an effort to counteract the naval blockade of Germany. The German *Kriegsmarine* was thus well-poised to add the U.S. to its list of targets.

Operation *Paukenschlag* (Drumbeat), a barrage of submarine attacks on American shipping, was launched in January of 1942, and soon Nazi U-boats were swarming up and down the East Coast, preying on American warships, tankers, tenders, and supply ships.

The German fleet was ready; the American military was not. The first few months of 1942 were basically a romp for the Germans, who referred to it as the "Second Happy Time" (the first being an earlier naval operation in which little resistance was offered). The British, who had been the victims of the first "Happy Time," recommended that ships avoid obvious standard routes, that navigational aids such as lighthouses be shuttered, and that a strict coastal blackout be enforced. Whether because of doubts about the soundness of the advice or because it was decided to allocate resources elsewhere, the suggestions of the British were by and large ignored. Although small shoreline communities were politely asked to "consider" calling for dim-outs, fear of negative effects on tourism and business shielded larger cities from such requests. All the U-boats had to do was watch for ships silhouetted against city lights at night.

After a few months, the Americans knuckled down and adopted more stringent measures. Blackouts were ordered and enforced, and my husband remembers being scolded, as a young child in Maine, for peeking under a cover draped over a radio to hide the light while the family listened to the news. Peacetime shipping lanes were abandoned and shipping was restricted to daytime in convoys, escorted by British corvettes specially designed for anti-submarine warfare. Unpredictably-timed daily patrols were implemented, and gradually the German "wolf-packs" retired to happier hunting grounds elsewhere in the Atlantic. Eventually German codes directing sub maneuvers were broken by the Allies, and Operation Drumbeat officially ended in July of 1943—although periodic submarine attacks on American ships continued until just a few days before Germany surrendered in 1945.

Still, although the Allies technically won the Battle of the

Atlantic by virtue of winning the War, the overall picture for America did not exactly smell like victory. Indeed, one historian claims that Operation Drumbeat "constituted a greater strategic setback for the Allied war effort than did the defeat at Pearl Harbor." More ships and more lives were lost in the former than in the latter, giving the lie to the popular notion that Pearl Harbor and Nine-Eleven were the only two successful strikes against the American homeland. Yet in all, only a handful of U-boats were destroyed.

* * *

So where was everybody while all this was happening?

A few people, of course, were of necessity informed about what was going on in their back yard. My father, in our oral history interview in 1979, told me with perhaps less than complete candor, "There was no attack on the East Coast, at least not at that time [of Pearl Harbor] —although there was some later." (A better interviewer than I would have pressed him on this issue.) His brother Sid, who was in the Coast Guard stationed in the Boston area, must have known. The "need to know" list, however, was apparently astonishingly small—reflecting the succinctly-stated policy of Ernest J. King, Chief of Naval Operations: "Don't tell them anything. When it's over, tell them who won." (This is the same Admiral King whose daughter once remarked that her father was very even-tempered: "He's always in a rage.")

Nonetheless, many individuals were uncomfortably aware that something was afoot, especially after blackouts were systematically mandated and enforced. A number of ships were torpedoed by U-boats within sight of New York and Boston. Other residents near the shore reported seeing eerie lights and other unnatural phenomena that were difficult to explain. Further news about the German presence trickled out when a couple of unsuccessful attempts to land German spies on American shores via U-boat were foiled.

On the whole, however, popular alarm was subdued by non-disclosure (with the rationale that news of sinkings so close to our shores might aid the enemy or undermine morale) and misdirection. No patriotic newsreels about these shoreline attacks were shown at American matinees (whereas German moviegoers regularly saw such footage). Citizens who had actually witnessed a U-boat attack, or the destruction of a U-boat, were asked not to reveal what they had seen. The famous slogan "Loose lips sink ships" was in fact coined to justify this veil of secrecy. The American people were never given a full account of the genuine danger they faced.

Partial knowledge and half-truths, of course, bred speculation and rumor—something that can be more dangerous than the truth. On Saturday, August 7, my grandmother noted in the journal she kept in the summer of 1943 during her annual stay on Cape Cod, "Gradually the news comes in with the story of the Busy Blimps. We hear a large convoy was going by and was attacked by a German submarine; that it has been sunk somewhere off our shore. We hear distant guns but never know whether they are real, or target practice somewhere." She drew pictures of her neighbor's clothesline on successive days, hoping to discern patterns in the size and spacing of the hanging towels that might, just might, be coded messages. Less amusingly, she also identified a potential German spy among the customers at a local bakery. "The authorities think someone is acting as a German agent along our shore. I think so too, but probably have my eye on the wrong person." Fortunately she was wise enough not to gossip about her suspicions; loose lips may sink ships, but false accusations can do their own kind of damage.

* * *

How close did Operation Drumbeat come to the Jersey Shore? Very close indeed.

Here as elsewhere along the Eastern Seaboard, residents periodically reported seeing Allied convoys being attacked by the U-boats and being awakened at night by the sound of explosions at sea. But the most shocking evidence came to light in 1991, when a fisherman's net caught on a massive object sixty miles off Point Pleasant, New Jersey. When a group of recreational divers made the 230-foot descent, they found an intact U-boat, along with its torpedoes and the remains of its crew. No records of attacks in the area or unaccounted-for U-boats could be found to identify the wreck.

Over the next six years, a team of professional divers led by John Chatterton and his partner Rich Kohler made it their mission to determine the identity of the wreck—dubbed "U-Who" because of the uncertainty. Their efforts were hampered by diving conditions so treacherous that they claimed the lives of three divers during the course of the exploration. (See Bernie Chowdhury's *The Last Dive: A Father and Son's Fatal Descent into the Ocean's Depths* [Harper Collins, 2000] for a heart-rending account of the deaths of Chris and Chrissy Rouse, an experienced father-and-son scuba diving duo.)

Chatterton and Kohler's first clue was the discovery of a knife inscribed "Horenburg," the name of a Radio Operator assigned to U-869, a Type IXC/40 U-boat. Their initial elation, however, was dampened by the news that U-869 had been sent to Africa and sunk off Casablanca on February 28, 1945 by an American destroyer and a French sub chaser. Reluctantly, the "Horenburg knife" was discounted. In 1997, however, serial numbers and other conclusive evidence were recovered confirming the identity of the wreck as U-869. Evidently the commander, Hellmut Neuerburg, had never received the orders diverting the sub to Gibraltar and instead perished off the Jersey Shore just a few miles from where I lived in Shark River Hills; I was two years old at the time.

U-Who continues to hold its secrets close—though not for want of ink spilled on the topic. Chatterton and Kohler, as documented in Robert Kurson's bestseller *Shadow Divers: The True Adventure of Two Americans Who Risked Everything to Solve One of the Last Mysteries of World War II* (Random House, 2005), concluded that the sub was probably sunk by one of its own acoustic torpedoes. Gary Gentile, another experienced wreck diver, hotly disputed this theory in his book *Shadow Divers Exposed: The Real Saga of the U-869* (Bellerophon Bookworks, 2006), citing logs from two destroyer escorts suggesting that they had sunk the sub and also arguing that the damage was more consistent with the destroyer attacks. The United States Coast Guard's official report, after a lengthy investigation, supported Gentile's conclusion, but Chatterton and his colleagues continue to believe that the two destroyers attacked the sub *after* it had been struck by its own torpedo. The truth may never be known.

* * *

Yet another piece of the U-Who puzzle was added when a German named Herbert Guschewski, after watching a preliminary version of a 2004 PBS NOVA episode about the wreck titled "Hitler's Lost Sub," approached the producers of the documentary. Guschewski had been the Second Radio Officer assigned to U-869 (and a close colleague of Martin Horenburg, whose knife had given the first hint about the sub's identity) but was hospitalized with pneumonia and pleurisy just before the boat departed and had thus been unable to accompany his crewmates on their first and only voyage. An interview covering his recollections of life on a U-boat and his feelings about being the sub's lone survivor is included in the final version of the NOVA program. It is worth watching.

THE RIGHT THING TO DO

my grandmother on the porch of her cottage, Shining Sands, in Wellfleet

World War II is perhaps the last war in American history to have had almost universal public support. It was not only the Great War, it was also the Good War, a war in which we felt clearly aligned against the forces of evil.

In his book *The Greatest Generation*, Tom Brokaw included within the scope of that term not only those who went off to fight on foreign soil but also those whose productivity at home made a decisive contribution to the war effort. It included not only my father-in-law, who spent most of the War as an Army Surgeon in Italy, but also his wife, left on her own to raise a child —my husband—who essentially never knew his father until he was six years old. It included my stepmother Rose, who put her career as a musician and a grade school teacher on hold to work as a welder. These were sacrifices made, and made willingly if not gladly, because Americans saw it as a collective commitment, not just a head count of troops sent by the President to fight and perhaps die in lands whose names they could barely pronounce. As Tom Brokaw put it, it was the "right thing to do."

* * *

My paternal grandparents had three sons of draft age. All served their country during World War II, but in surprisingly different ways – a testament, perhaps, to the family ethos of independent thinking:

King, the eldest, worked for the Army as a civilian scientist. There was never any question of his being sent overseas because his work in developing radar was considered essential to the war effort and in the event proved crucial to the Allied victory.

Sid, the youngest of the Stodola brothers, joined the Coast Guard in October of 1942 and served until February of 1946. Shortly after Pearl Harbor and the declaration of war with Germany and Japan, the Germans launched Operation Drumbeat, a phase of the Battle of the Atlantic, as described above, in which German U-Boats attacked merchant shipping and naval vessels up and down the eastern coast of the United States (a situation for which the American military, despite the Coast Guard's famous motto *Semper Paratus*—"always ready"—was at first woefully unprepared). In addition to the need to protect shipping, there was also considerable concern about the possibility that the Germans might be able to establish an enemy presence on land. Possibly the prominence of this activity in his own back-yard inspired Sid's decision to enlist in the Coast Guard. Of course, the Coast Guard also played a prominent role in naval warfare worldwide. Did Sid serve overseas? I do not know.

Quentin, the middle son, chose the most controversial path (even within his own family), registering as a Conscientious Objector—not a popular position especially during World War II, and even more especially for someone who was not a Quaker.

* * *

I decided to delve a little deeper into the contribution of these three young men to the war effort, and to invite their mother, Beatrice King Stodola, to help me out. Almost as though she were anticipating my queries, my grandmother was prescient enough to keep a journal titled *Saint Barbara's Hill: A Cape Cod War-time Log*, in which she documented her joys and (many!) anxieties, as well as the comings and goings of her "boys," during her annual stay at Shining Sands, the family's beloved vacation cottage, in the summer of 1943. (My poor grandfather, except for an occasional stolen weekend getaway and a longer stay later in the summer, remained sweltering at his office job in New York City.)

I feel particularly blessed to have this journal because it gives me a glimpse of the adult relationship I might have had with my grandmother had she lived long enough. For example, on Thursday, August 5, she wrote, "My husband swims every day, but [I] must confess I only go in when I feel like it, which is not very often." I'm with her; I'd enjoy swimming so much more if only I didn't have to get wet! And again, a few days later, on Friday, August 13: "I had rather a disturbing letter from one of my sons [Quentin]. He may not bring his wife with him when he comes on furlough. It's just too bad! Received a letter from the wife of another son [King] with a picture of my grandchild—bless her. It was so nice to have daughters when my sons married." As the mother of daughters, I feel exactly the same way, *mutatis mu-*

tandis, about my two sons-in-law!

The journal offers fascinating insights into life in the 1940s in a country at war. My grandmother repeatedly refers to the nightly blackouts and bemoans her inability to get enough butter to make a blueberry pie. More chilling is the pervasive sense of fear and suspicion that permeates the journal, compounded by the inability to know exactly what's going on: Is that apparently harmless shopkeeper an enemy agent in disguise? Is that man on the beach really just burning his garbage, or is he sending signals to the Enemy? Although these concerns may seem exaggerated, we now know that German threats to coastal shipping were much more extensive than previously believed.

Despite her ongoing concern about German submarines and suspicious strangers, however, she seemed to be in greater danger from target practice conducted by the American military on the nearby dunes. The serenity of her vacation retreat was incessantly violated by the sound of the "ack-acks" in the (not-so-distant) distance—accounting for the title of her journal: "We shall have to change the name of this hill to Saint Barbara's Hill. She's the patron saint of gunners I am told." During one of his weekend visits, my grandfather retreated several times one afternoon from the task of putting up screens on the second-floor windows because the tracer bullets aimed at the target (towed by an airplane!) were just too close for comfort.

But the most obvious impact on her family, of course, was the involvement of all three of her sons in the war effort. They are here "only in spirit, just now," she notes. "'Uncle Sam' keeps them busy at the moment."

My father, working overtime at Camp Evans and spending what little spare time he had fulfilling his role as a new father, evidently never made it to the Cape at all that summer. My mother, however, kept the lines of communication open as women so often did (and do). On Thursday, July 29, my grandmother wrote, "Pictures came today of our first grandchild [that would be me]—five months old. We have all been wondering whom she looked like but these snap-shots certainly look as her dad, my oldest son, did when he was a baby."

My Uncle Sid, a carefree bachelor, seems to have shown up whenever he had a few days' leave. On Saturday, July 24, she wrote, "About noon a telegram came from our Coast Guard son saying he would be in this afternoon between three and four—probably a forty-eight-hour leave. He arrived on the four o-clock bus from Boston. The rest of the day was spent in TALK and EATS.... After we pulled the black-out curtains we played three-handed bridge, in the midst of which the mother cat walked in proudly carrying a half-grown rabbit. The little thing seemed partly alive so the Coast Guard rescued it and put it out in the silver leaf." And then the next day, "We went out early this morning to see if the rabbit had escaped safely. It had! Somehow saving the life of a rabbit, and guns on the next hill, seems rather incongruous.... The Coast Guardsman left on the late bus very proud of his new stripe—he is now a seaman first class." And again, on Sunday, August 8: "We sat on the edge of the dune with [Sid] and watched groups of P.T. boats on the horizon.... [He] left at 7:30 with a package of cigarettes given him by the owner of the nearby village store. Those little things mean so much more to the boys than anyone realizes."

relaxing with Sid on the beach

About my Uncle Quentin, whom I came to know very well when I was an adult and whom I would describe as a perfect blend of severity and saintliness, I will leave the last word to my grandmother, who got it exactly right in her comments on Monday, August 16: "[Quentin] is leaving this week-end to work for the Government in New Mexico. He has been at a C.P.S. [Civilian Public Service, which provided American Conscientious Objectors with an alternative to military service during World War II] camp in New Hampshire since shortly after the war began. I stand by all of my sons like the Rock of Gibraltar, whatever their beliefs. Who am I to say what is right or wrong when it comes to creed or religion? He has lost eighteen pounds being a human guinea pig in a diet experiment for the Harvard Fatigue Laboratory. They are trying to find the best food to send to the starving millions abroad. Last year he was a guinea pig for a louse experiment conducted at the camp by the Rockefeller Institute. In the last war thousands of boys lost their lives through disease spread by lice. The powders developed in this experiment came in time to be used by our boys in Africa."

"Getting our lousy underwear at the Lyceum."

I'm proud of all the Stodola boys, for their courage in forging their destinies and for their signal contributions to the war effort. Each in his own way did work that saved countless lives. How many families are vouchsafed such a privilege?

ROSIE

our very own Rosie—Rose Balaban Lamay Stodola (1921-2017)

I n late 1940, against the backdrop of a long-simmering war that was starting to boil over, President Franklin D. Roosevelt was preparing for his upcoming State of the Union address, struggling to find a way to express his longstanding conviction that the response to any world crisis, whatever it might be, should not be simply an expression of fear. ("The only thing we have to fear is fear itself," he had famously said in 1932, speaking not of war but of the Great Depression.)

Then, on January 1, 1941, while crafting the third draft of his message to Congress, Roosevelt was struck by what he believed to be an inspired way of encapsulating his vision for both America and the world—that is, by enumerating what he called "The Four Freedoms": freedom of speech, freedom of worship, freedom from want, and freedom from fear. The peroration he dictated to his secretary at that moment was retained almost verbatim in the actual speech five days later.

How disappointed Roosevelt must have been, then, when the major American newspapers, while covering the address in detail, pretty much ignored the four freedoms passage. Even after the U.S. entered the war in December of 1941, polling results showed that although 80% of Americans responded favorably to the underlying ideals, fewer than 25% could name even one of Roosevelt's four freedoms, and 61% had never even heard of the four freedoms at all. As a catchy slogan Americans could unite behind, the four freedoms were a nonstarter.

And then, almost miraculously, the four freedoms were rescued from oblivion by the painter Norman Rockwell, who, wanting to do his part to support the war effort, retired to his studio in Arlington, Vermont in 1942 and in the space of seven months translated the four freedoms into four compelling images. Although the subject matter was taken from the everyday life of small-town America, they were in many ways aspirational rather than realistic. As Rockwell knew well, not all Americans enjoyed all four freedoms and some enjoyed none. Nonetheless, the scenes were recognizable to all, and the four freedoms, so abstract in Roosevelt's words, became memorable in Rockwell's hands.

These beloved paintings appeared as covers on four consecutive weeks of the *Saturday Evening Post* in February and March of 1943, as well as on posters issued by the U.S. Government Printing Office and on postage stamps. A photograph in the Camp Atterbury, Indiana Archives dated April 12. 1943, showing a WAC and a soldier flanking a four-freedoms poster, suggests that they were widely distributed on military bases (likely including

Camp Evans).

* * *

Back before the Covid-19 pandemic transformed our cultural lives, perhaps forever, we visited a superb traveling exhibit at the Henry Ford Museum in Dearborn, Michigan called Enduring Ideals: Rockwell, Roosevelt, and the Four Freedoms. Although the *Four Freedoms* paintings were the centerpiece of the exhibit, it encompassed much, much more, covering Rockwell's growing commitment to contributing to the war effort in a meaningful way, and later to the advancement of civil rights. The work of other artists and writers also appeared, along with photographs and an evocative assortment of memorabilia.

As I wandered through the aisles, I felt sure there was something to be said about Norman Rockwell in the context of our daily lives on the Jersey Shore. After all, his fame as an artist/illustrator peaked between 1941-1946. My parents subscribed to the *Saturday Evening Post*, for which Rockwell created more than 300 cover illustrations over the course of his 47-year association with the publication. I remember looking forward to each issue as soon as I was old enough to read, always hoping Norman Rockwell's instantly recognizable work would be featured on the cover, since I enjoyed both the visual humor and the satisfying sensation of "getting it." And yet I can't recall any specific discussions of Norman Rockwell, and when I polled my siblings, neither could they. I can only conclude that Norman Rockwell was so much part of the water we swam in, the air we breathed, that his work was too familiar and omnipresent to be worthy of comment.

The *Four Freedoms* were in some ways a turning point in Rockwell's work, transforming him from an illustrator for whom the war effort was simply something he had to capture as part of his weekly deliverable to a painter with a much more ideological commitment and a will to determine how he could use his special gifts to contribute to that effort. A fire in his studio at around that time destroying all his irreplaceable costumes and props provided him with an additional impetus to focus on contemporary issues.

One outcome of all this introspection was Rockwell's focus on a series of "characters" who came to stand for the American "can do" response to wartime mobilization. Perhaps the most notable was Private Willie Gillis, who appeared on several Post covers. Another, painted not long after the *Four Freedoms* tetralogy, was Rosie the Riveter, who appeared on the cover of the *Saturday Evening Post* on May 29, 1943.

* * *

I always loved the idea of Rosie because of what she represented—a woman who, when the men were called away to fight a war, willingly did the work of the world, supplying both wartime needs and what was needed on the home front, and was paid for it. But no matter how often she was admonished that once the war was over she would have to cede her job to a returning serviceman who needed the work to support his family, that genie could never quite be put back in the bottle, and Rosie unexpectedly turned into an agent of radical social change.

I also felt a special affinity with Rosie for a more personal reason: In 1968, my widowed father married a woman who had actually been a Rosie. My stepmother Rose (yes, that was really her name) was a gifted pianist who as a young woman had performed in Carnegie Hall; but when the war broke out she went to work as a welder at Grumman Aircraft on Long Island. After the war she resumed her musical career, working as a music teacher in the Suffolk County School System, but she always enjoyed talking about her service as a real live "Rosie." Rose was a kind-hearted and good-humored woman, not unlike my mother; she nursed my father throughout his final years as he sank deeper and deeper into the fog of Alzheimer's, and for this I shall always be grateful.

So imagine my surprise, as a self-professed Rosie buff, to learn at the Rockwell exhibit that 1) Norman Rockwell had painted an enormously popular picture of Rosie the Riveter, with which I was completely unfamiliar; and 2) the image I and most others knew as Rosie was not the one Rockwell had painted and was possibly not even Rosie.

How could this be?

The answer is that "Rosie the Riveter" actually started out as a song, written in 1942 by Redd Evans and John Jacob Loeb. There was no real-life Rosie; or perhaps more accurately there were thousands. As Robert Lissauer, a business partner of Loeb's, later recounted, "They wanted to write a song about women who were working for the war effort for the country. So they just made up the name 'Rosie the Riveter.' You pick a name for the alliteration and you go ahead and write it." The song was recorded by numerous artists; one, by the big band leader Kay Kyser, became a national hit.

The song inspired a number of paintings, of which Rockwell's was the most popular and widely known at the time, but far from the only one. Rockwell's Rosie wore a blue jumpsuit, with a rivet gun in her lap, a sandwich in her hand, and a copy of *Mein Kampf* beneath her foot. Her lunchbox was labeled "Rosie," a backlink to the popular song by Evans and Loeb. Rockwell's Rosie had red hair and was so hefty and muscular that Rockwell felt he had to apologize to his neighbor Mary Doyle, a much more petite woman who had served as his model.

Among the many other contemporary Rosie and Rosie-ish images was the one with which we are most familiar today, a determined woman with her dark hair swept back by a polka-dot bandanna and flexing her biceps, with a text balloon asserting, "We Can Do It!" This image, painted by a young artist named J. Howard Miller, was commissioned by Westinghouse Electric Corporation in 1943 as part of a series of posters boosting support for the war effort on the home front. It was displayed for a couple of weeks in factories in Pennsylvania and the Midwest, then replaced by the next-up poster. Fewer than 1,800 copies were printed.

It is not clear that the woman depicted In Miller's poster was ever intended to represent anything other than a generic woman who should be welcomed—*temporarily*—into the workforce, an idea that met with considerable opposition despite the obvious need for many extra pairs of hands. Nothing on the poster identifies the woman as Rosie or hints that her job was riveting (as opposed to welding or even mopping floors) —though the popularity of the Evans and Loeb song probably made Rosie spring to the minds of many viewers.

* * *

And there things stood for decades, until circumstances conspired to create a need for a heroine like Rosie. The 1980s marked the start of 40th-anniversary celebrations of World War II (including my father's largely unsuccessful attempt to commemorate Project Diana in 1986). It was also the time in which the second wave of feminism was winding down, having for the most part (with the glaring exception of its failure to ratify the Equal Rights Amendment) met its goals, and many women were encouraged by the legislative and social victories their efforts had brought about. The National Archives, faced with the budget cuts of the Reagan era and looking for a way to generate income by capitalizing on both World War II nostalgia and the feminist wave, somehow hit upon the Miller image, licensed it, and plastered it on tee-shirts, mugs, and other souvenirs. It turned out to be an inspired choice.

Although licensing Rockwell's *Rosie* would have been a much more expensive proposition, there are probably additional reasons why his painting was passed over for this campaign. Miller's portrait, more feminine than Rockwell's, was less likely to create uneasiness around the issue of gender-bending. The Miller painting is also less of a period piece; "We Can Do It," though at the time it implied "win the war," translates more fluidly to the hope of achieving other goals than does the symbolism of Rosie tromping on *Mein Kampf*. The choice of Miller's painting, far from being obvious, was a stroke of genius. Its moment had arrived.

What is harder to fathom is the alchemy by which Miller's painting came to occupy Rosie's identity, probably displacing Rockwell's *Rosie* for all time.

✳ ✳ ✳

Among my faves from the Four Freedoms exhibit were two Rockwell paintings more or less bookending the War era that appealed to me for their engaging portrayal of youth. *The Marble Champion,* dated 1939, shows a girl ready not only to compete with the boys but to win; with her red hair and resolute expression, she could have been a young Rosie the Riveter. *Back to Civvies* depicts a returning soldier ruefully noticing that the hems of his old trousers barely reach his anklebones—a poignant reminder of how very young were the still-growing boys sent off to do battle on faraway shores.

DEFEATING THE "DIVINE WIND"

I n 1274 AD and again seven years later, Mongol fleets led by Kublai Khan launched major attacks on Japan. On both occasions, according to legend, massive typhoons destroyed the vessels and foiled the invasions. The Japanese believed these storms had been sent by the gods to protect them from conquest and called them the "divine wind."

Or in Japanese, *kamikaze*.

In the waning days of World War II, when Japan's defeat was all but inevitable but surrender was unacceptable, Emperor Hirohito "asked" Japanese pilots to become kamikazes, divine winds again defending the homeland by deliberately crashing into Allied warships. Preference for death over defeat or capture was deeply embedded in the Japanese military culture, as was the tradition of absolute loyalty to the Emperor, the gods' representative on earth. The number of volunteers exceeded available aircraft, and extra men were sometimes sent to accompany the official pilot, perhaps to provide moral support.

Some kamikaze aircraft were fashioned from existing planes, others were purpose-built—most notoriously the MXY-7 "Ohka", which was actually designed to kill its pilot. (Although the Americans sometimes dismissed these pilot-guided missiles as *Baka* bombs—a Japanese pejorative roughly meaning "stupid"—

it would probably be more accurate to describe them as the original "smart bombs.") The first kamikaze mission struck in late October of 1944. In the end, nearly 4,000 kamikaze pilots died and more than 300 managed to hit a ship. Reportedly, more than 70 U.S. vessels—aircraft carriers were a favored target—were sunk or damaged beyond repair.

Most soldiers go into battle understanding they may be called upon to make the supreme sacrifice. As we know from even more recent history, however, deliberate suicide attacks present special challenges since no exit plan is required. Kamikaze pilots were instructed to fly low over the water and to keep the mountains at their backs. This enabled them to evade Allied radar, which aimed its beams away from the ground to avoid signals from non-moving objects at a fixed distance from the antenna —producing "clutter" on the display that made detection of moving targets more difficult. Essentially, the Japanese military had found a radar blind spot they could exploit, for example, through suicide missions that could remain "under the radar" until they self-destructed.

Urgent appeals were made to the radar laboratories of the Army, the Navy, Bell Laboratories, and the Massachusetts Institute of Technology to find a way to eliminate the blind spot. At the time, my father was head of the Special Developments Group, which had produced many radar improvements including the Army's first operational moving target radar. Based on his earlier work on the Army radar series, the SCR-270/271 radars, which had successfully detected Japanese planes approaching Pearl Harbor in 1941 (information that was unfortunately discounted by the commanding officer on duty), he was assigned to lead a team of experts in responding to the call for improved detection of moving targets low on the horizon by filtering out background clutter.

To accomplish this goal, within the context of very limited time and budget, the team made major modifications to the SCR-270 to stabilize emissions and improve detection of small frequency shifts (Doppler effect). The large return signals produced by stationary objects (i.e., background clutter) were then processed using small time constants, resulting in rapid decay on the radar screen, while the smaller Doppler-shift signals were processed using longer time constants to obtain greater persistence of the moving target on the screen, producing a distinctive "writhing" pattern that could be readily perceived by the human eye.

These modifications were carried out onsite by available government personnel—no time to job it out!—and then field-tested both in a mountainous region near Ellenville, New York and on Navy landing craft. The updated equipment performed beautifully. Gone was the blind spot; kamikaze pilots could no longer fly under the radar.

Had it not been for Project Diana, my father's work in thwarting the kamikaze strikes would probably be the accomplishment for which he is best known today.

Less than a year later, on August 6, 1945, the U.S. dropped an atomic bomb on the Japanese city of Hiroshima; on August 9, another was dropped on Nagasaki. The calculus that went into this decision—the number of Allied lives presumably saved by ending the war sooner rather than later, and whether the same goal might have been achieved without targeting large civilian populations—will undoubtedly be debated as long as it is remembered. In any event, the Japanese surrender was announced less than a week later, and the official documents were signed on September 2, 1945. For many historians, however, the verdict remains that although the A-bomb may have ended the war, it was radar that **won** the war.

My father later wrote that this "earlier work on moving target detection had prepared us well for [Project Diana]." And indeed, the approach he and his staff brought to the task of making kamikaze flights visible—work rapidly but carefully, using materials already on hand—foreshadowed his approach to bouncing radar off a very large target moving through space and, in little, to household repairs, where clever jury-rigging was elevated to a fine art.

D-DAY

"Full victory—Nothing else." Those were General Eisenhower's orders to the paratroopers of the 101st Airborne Division at the Royal Air Force base in Greenham Common, England, as they waited to board their planes for the first assault wave.

Lt. Col. Simon R Sinnreich, America's highest ranking Jewish officer during World War II, had a similar take. Our family didn't know him during that era, but Si and his wife Emilie became close friends of my parents soon after we moved to Long Island in 1956—a friendship that lasted to the end of their lives and has continued into the next generation of Sinnreichs and Stodolas. Although it wasn't easy to get Si to talk about the war or his service in Europe, my husband, who has a gift for drawing people out, once asked him how many troops the generals were prepared to lose in the Battle of Normandy. Si's answer was stark and simple: "As many as it took."

For most of us, the term "D-Day" evokes haunting images of wave after wave of landing craft approaching the Normandy beaches, of the hapless paratrooper left hanging for hours after his chute became entangled on a church steeple in the nearby village of Sainte-Mère-Église, and of course of the Normandy American Cemetery with its seemingly endless expanse of crosses. On the 75th anniversary of D-Day in June of 2019, with solemn military ceremonies and moving first-person accounts by the few remaining veterans, now in their nineties, flooding the media, these heroic exploits and appalling sacrifices claimed most of our attention.

Less well known, though also worthy of recognition, are the contributions of the so-called "wizard war" to this historic victory, and the participation of Camp Evans in this effort.

To induce Field Marshal Rommel to hesitate or possibly even to deflect the German Panzer Divisions to the wrong place, Eisenhower stationed a shadow invasion fleet in Northern England, complete with dummy inflatable tanks, and leaked misleading information through his network of spies to suggest the invasion would take place at Calais, not Normandy. He knew, however, that this ruse would not be enough to deceive Rommel, because the Nazis had radar units all along the French coast searching for signs of an invading fleet. The Allies' only hope of evading these tireless sentries was to destroy as many of them as possible and then to use the same strategy of adding competing information to the mix.

The radar scientists at Camp Evans, along with their counterparts in Great Britain, the U.S. Navy, Harvard, and the Massachusetts Institute of Technology, were tasked with developing the equipment needed to carry out these plans. Their efforts enabled bombers to zero in on German radar sites, to interfere with (jam) their communications, and to introduce confusion by dropping "chaff" —mostly strips of aluminum—to create a cloud of indecipherable images on Nazi radar screens, ploys that caused Rommel to delay sending Panzer Divisions to Normandy long enough for the Allies to establish a beachhead. As a result, the German air response was next to non-existent. In addition, radar sets designed at Camp Evans landed on the beaches to protect the troops as they fought to fend off Panzer attacks.

By D-Day, radar and its military uses had clearly come a long way in both sophistication and precision since the Chain Home network, a series of radar antennas strategically placed all along the British coastline to detect and track incoming aircraft. Using radar not only to obtain information but to spread disinformation, then called radar countermeasures, is now known as Electronic Warfare or EW, the field in which my father continued for the remainder of his career.

B. CAMP EVANS

THE TINY SHARK RIVER TAKES ITS PLACE IN HISTORY

S hark River Hills was originally developed as a seasonal resort community on a 700-acre parcel of beautiful rolling hills surrounded by the Shark River. During World War II, however, it attracted many young men stationed at Camp Evans who saw it as a good place to raise their families, just a short drive or bike ride from work, and who made it their full-time home. These two communities overlapped but never completely merged, and the houses next door to and across the street from ours were shuttered and empty most of the year. The annual return of the Wardell children, after school let out in Flushing, wherever that was, was an eagerly awaited event (with the added element of surprise, since we never knew exactly when their car would reappear in the driveway), and their departure in late August much lamented.

For both full-timers like us and part-timers like the Wardells, the Shark River provided the focus of community activity in the 1940s. I practiced riding my bike no-handed down the impossibly steep (or so it then seemed) Riverside Drive hill; then, if I had a few pennies in my pocket and didn't want to ride all the way to the Cracker Barrel for a larger selection, I stopped at a little ice cream stand on the beach, now the site of a large marina. I didn't really want to swim in the river itself (because of the fierce-looking but harmless horseshoe crabs and the smaller but more aggressive crabs that given half a chance would chomp your toes), though many people did; I much preferred the ocean beaches at Avon, Belmar, and Ocean Grove, with their roaring surf and vast expanses of white sand. But I loved beachcombing along the river's edge for shells and starfish and whatever other flotsam and jetsam the tides washed up. I loved building riverside sandcastles and once in a while—though I rarely stopped my restless gamboling and cartwheeling for long enough to do it —basking in the warmth of the summer sun.

How the Shark River got its name is lost in the mists of time, though fanciful explanations abound. Known to the Lenni-Lenape Algonquins as the Nolletquesset, it is listed on 18th-century maps of the White settlers as either the Shirk River or the Shack River. Possibly its various similar names were somehow conflated, or possibly its crescent shape reminded someone of a shark. One legend had it that a large shark was swept up the river to its death in the mid-1800s. At any rate, with the dubious exception of that lone unfortunate, sharks haven't lived there at least since the Middle Miocene period, though fossil shark teeth can still be found along its banks. Certainly the possibility of a shark attack wasn't high on anyone's worry list.

Because the Shark River was the only river I knew, and because I knew it so well, it formed my entire concept of what a river is. It wasn't till I reached adulthood that I learned that the Nile, the Amazon, and in fact all the other rivers of the world were not salty like the Shark River—leading me to infer that the lowly little Shark, less than 12 miles long, was salty because of ocean backwash. Much later still—just a few years ago, in fact—I learned that the Shark River is not technically a river at all but rather a tidal basin or small bay of approximately 800 acres, which accounts for its saltiness without need of my "ocean backwash" theory.

So: No sharks and not a river. Who knew?

* * *

Fortunately for the future of communications, Guglielmo Marconi made it his business to know. After several months of surveying the entire eastern seaboard in 1913, he selected a hilltop on the south bank of the Shark River basin as the perfect spot for the Belmar Receiving Station, operated by the newly-formed Marconi Wireless Telegraph Company as the world's first permanent transoceanic communications center. It was located far enough from the shore and high enough to be protected from Atlantic storms. In addition, the elevation provided a low take-off angle for radio emissions, and the long-wave frequencies used at the time benefited from salt water reflection (i.e., the ocean as a "virtual antenna") for beaming signals to Europe.

The original buildings (starting with a hotel to house the workers) were completed in 1914, serving as a relay station for the transatlantic transmitting facility at New Brunswick, New Jersey, part of the New York-to-London link of the "World Encircling Wireless Girdle." According to a contemporary report, the station included "a mile-long bronze-wire receiving antenna strung on six 400 foot tall masts with three 150 foot balancing towers along the Shark River." Morse code messages traveled via telegraph landlines to and from the Belmar Station, the New Brunswick transmitter, and an office in New York City.

When the U.S. entered World War I, the station was commandeered by the Navy to handle transatlantic radio traffic, a task in which it admirably acquitted itself. Following the armistice, all of Marconi's assets in the U.S., including the Belmar Station and its equipment, were transferred to the Radio Corporation of America, an entity newly formed for the purpose of keeping American radio under American control and ownership. Unfortunately, RCA abandoned the Belmar Station in 1924, and the site, now fallen on hard times, was used as a meeting place for organizations as disparate as the Ku Klux Klan and the King's College, a theological school.

Then, even before Pearl Harbor, the U.S. Army Signal Corps began a search for a new research facility. Like Marconi, the Army's communications experts scoured the eastern coastline for an ideal site. The same features that made the Shark River basin favorable for radio propagation also lent themselves to the study and testing of radar, allowing ships and decoys at sea to be used to determine accuracy and range. And although much of the original Marconi equipment had disintegrated or been removed by the time the Army's advance scouts arrived on the scene, its history as a locus of communications know-how and research must have favorably disposed them to the site.

In 1941 the property was purchased by Wall Township for the Army's use as the Signal Corps Radar Laboratory—renamed the Evans Signal Laboratory and subsequently, at its official dedication in 1942, the Camp Evans Signal Laboratory. Dozens of additional buildings were quickly built (for a total of around 134 by the end of the war). Top-secret research carried out throughout World War II played a central role in the development of radar as an effective weapon for detecting kamikazes, analysis of captured German and Japanese radar, friend-or-foe identification, and other vital wartime applications. We may never know the full story of the research carried out at this facility, but it has been said that had the Axis powers known, Camp Evans would have been one of their prime targets.

By the end of the War, the site was perfectly poised to make the leap to space communications. The world was ready for Project Diana, and Project Diana was ready for the world.

CAMP EVANS AND THE LEGACY OF BUCKMINSTER FULLER

five Dymaxion Deployment Units awaiting restoration (August 2017)

Along with the Beatles, Ravi Shankar, Marshall McLuhan, Woodstock, and psychedelia, Richard Buckminster "Bucky" Fuller (1885-1983) was part of the interior landscape of my generation. It was this eccentric and peculiarly American genius—architect, designer, engineer, planner, visionary, futurist—who popularized the geodesic dome, a distinctive hemispherical shape composed of triangular surfaces that distribute structural stress to maximize strength and stability. This remarkably efficient structure was the darling of not one but two World's Fairs, in New York in 1964 and Montreal in 1967.

It wasn't until I heard about his improbable link with radar research at Camp Evans, however, that I learned of his earlier work, before the geodesic dome catapulted him to fame. In the late 1920s, Fuller unveiled the first of many iterations of a new approach to home design, the Dymaxion house, launching a line of inventions associated with the Dymaxion concept. The term "Dymaxion" was coined by Fuller in collaboration with Waldo Warren, an adman hired by Marshall Field's in Chicago to help promote a display model of the Dymaxion house. A portmanteau word created by jamming together the words "DYnamic," "MAXimum," and "tensION," it was intended to convey Fuller's underlying vision of achieving a "maximum gain of advantage from minimal energy input."

Fuller was so taken with the term that he applied it not only to the Dymaxion house but also to the 3-wheeled Dymaxion car (potentially a flying machine as well), the molded plastic Dymaxion bathroom, a unique world map projection known as the Dymaxion map—and the main topic of this essay, the Dymaxion Deployment Unit. He renamed his journal the *Dymaxion Chronofile*. He advocated, and actually practiced for a couple of years, what he called Dymaxion sleep, a polyphasic sleep schedule that involved taking 30-minute naps whenever he became tired—generally after around six hours—allowing him to function with considerably less sleep than today's sleep gurus insist upon. He stopped not because of any ill health effects but because he found it put him totally out of phase with the rest of the world.

Fuller's goal in developing the Dymaxion house was to produce an inexpensive, mass-produced modular dwelling that could be airlifted to its final location in kit form and easily assembled on the spot. The original versions were hexagonal in shape, with a domelike roof and a central supporting mast allowing them to be raised off the ground by one story. The Dymaxion house was ahead of its time—way ahead—in employing photovoltaic cells, wind generators, micro-hydroelectric systems, pumps for onsite water purification, and packaging toilets that deftly shrink-wrapped solid waste—all in support of off-the-grid living at a time when the virtues of such green innovations were apparent to few besides Fuller.

Only two Dymaxion houses (having by then evolved into circular Dymaxion Dwelling Machines) were actually built, the Barwise (in 1945) and the Danbury (1946). A third, hybrid version, the Wichita House, was assembled by an investor named William Graham from salvaged parts of the two prototypes, which he purchased in 1948, and was occupied by his family for 30 years—although as an accessory to an existing house rather than as a freestanding living unit. In 1990 the house and all the leftover Dymaxion house parts were donated by the Graham family to the Henry Ford Museum in Dearborn, Michigan, which lovingly restored it and placed it on display in 2001.

Meanwhile, the Dymaxion Deployment Unit or DDU, a considerably less elaborate version of the Dymaxion house, was created in 1940-41, initially for British military use. The inspiration for these buildings came during a drive through the Illinois countryside, where Fuller became fascinated with the corrugated metal grain bins on the farmsteads that dotted the landscape. Adapting the construction principles exemplified by the Dymaxion house, he designed a structure with circular walls, ten porthole windows, and a dome-like roof, with a down-draft ventilator at the apex to provide natural air conditioning. Unlike the Dymaxion house, the DDU was designed to sit directly on the ground.

Although DDUs were originally intended as prefabricated bombproof shelters, Fuller envisioned multiple other wartime and peacetime uses including as storage space, small workshops, and (in the form of two side-by-side structures) vacation cottages. ("How to be Comfortable though Bombed," read one headline to a feature on the units; "A Shelter in War—A Beach House in Peacetime," proclaimed another.)

In 1942, the Army Signal Corps commissioned 200 DDUs for use around the world. The Butler Manufacturing Company of Kansas City, Missouri, manufacturer of the silos Fuller so admired, was engaged to produce the units, which sold for $1,250 apiece. Probably 100 or so were actually built, until production was finally halted by the wartime shortage of steel.

Perhaps driven by the demands of wartime and the simplified design, the DDU, unlike the Dymaxion house or the Dymaxion car, was the only Dymaxion concept that ever came close to being mass-produced—if numbers in the double or low-triple digits can be called "mass-produced." Yet ironically, it was in some ways more Dymaxion in name than in fact, having at least as much in common with the corrugated sheet metal grain bins it was modeled on as with the much more elaborate and sophisticated Dymaxion house. Still, it filled a wartime need for cheap, sturdy, easily assembled portable structures that could be used as houses, emergency shelters, and specialized workspaces, and it might have enjoyed even greater success had continued production been permitted. (A copy of the Army's instructions for assembling a DDU is preserved in the Ft. Monmouth Command Historian's collection.)

Between 1941 and 1943, around 28 DDU's were ordered and installed on circular concrete pads at Camp Evans for use by the Signal Corps' radar research program. Parts for the AN/TPS3 radar unit, invented by Dr. John Marchetti to fill an urgent need for lightweight, transportable early warning devices by troops recapturing islands in the Pacific Theatre of Operations, were stored in five of the buildings. There is also evidence that DDUs were used as protected areas for workers conducting potentially flammable experiments or handling hazardous materials.

A shout-out here to Fred Carl, local history buff extraordinaire and founder of the InfoAge Science/History Learning Center on the old Camp Evans site. In 1985, Fred acquired a property adjacent to Camp Evans and soon became obsessed with the place and its enigmatic past. As his research brought more and more information to light, he worked closely with the National Trust for Historic Preservation to protect Camp Evans and its treasures.

I have no idea what Fred actually said when he first encountered these odd yurt-like structures, but it must have been something along the lines of "What the heck are these?" It wasn't until 1996 that the mystery was finally solved, with the completion of a historic resources study conducted by the Department of Defense. Without Fred's heroic efforts, this historic legacy of Fuller's work would likely have been leveled by now.

When the old Camp Evans campus was divided between InfoAge and the Wall branch of Brookdale Community College, the DDUs that were on the Brookdale property were disposed of. Although only a fraction of the original structures remain (a 2013 *New York Times* article spoke of 12, but my husband and I counted only 11 in 2017), the InfoAge collection is still probably the largest assemblage of DDUs anywhere in the world and perhaps the best surviving example of the Dymaxion phase of Fuller's work.

When we visited in 2017, the grounds of InfoAge were largely frozen in time, strewn with a ghostly assortment of mobile radar units, jeeps, and outbuildings ranging from the merely quirky to the downright bizarre. Making a virtue of necessity, for several years InfoAge embraced its own spookiness by turning the run-up to Hallowe'en into a month-long fear-fest called Camp Evans Base of Terror (CEBOT). CEBOT turned into a major fund-raiser for InfoAge, attracting dozens of teenagers on chilly October evenings to experience the thrill of being scared out of their wits, and these mysterious ruined structures were the stars of the show.

CEBOT, originally intended to tide InfoAge over till other, perhaps more dignified sources of funding were secured, is now gone forever, its ending only slightly hastened by the Covid-19 pandemic. Ironically, dispensing with the need for spookiness and Hallowe'en props has resulted in a major landscaping up-grade; as CEO Mike Ruane declares, "the campus has never looked better!" Work on scraping and painting some of the DDUs is now underway; the interior of one will be fully restored to its World War II condition to serve as an exhibit.

A fitting tribute—to both Bucky Fuller and Project Diana.

HOMECOMING

the old Marconi Hotel, now a part of the InfoAge Science Museum Complex

O n Saturday, August 26, 2017, my father was posthumously inducted into the "Wall of Honor" at the InfoAge Science History Learning Center, site of the former Camp Evans, for his contributions to long-range radar development during World War II and his leadership role in Project Diana, the first successful attempt to bounce radio waves off the moon.

Among the crowd gathered at InfoAge for the banquet and

awards ceremony were my three siblings, my husband, my sister-in-law, and my niece and her family (including my grand-nephew, the youngest and cutest attendee). My brother gave a brief speech accepting the honor on behalf of the family and acknowledging the beautiful plaque to be mounted on the Wall along with Marconi and other Camp Evans notables. It was a moving experience that left much of the audience in tears.

* * *

My father would have felt not only deeply honored, but also, in a sense, vindicated. The Army, embarrassed by being caught somewhat off guard by the feat and by early wild press speculations on the significance of the Project Diana experiment (radio control of "space ships"! mapping the surface of the moon! verifying Einstein's light deflection theory!), adopted a deliberate policy of curtailing further publicity on Project Diana. This meant not only keeping the press at bay but also limiting what publicity it permitted to only those men who remained in its employ, effectively excluding all five engineers who formed the core Project Diana team. Jack DeWitt, King Stodola, Jack Mofensen, and Herbert Kauffman had all moved to private industry shortly afterward, in part to continue pursuing their research careers, since the Army was now jobbing out much of the research that had originally attracted them to Camp Evans. Harold Webb went on to an academic career during which he continued to carry out research on issues raised by Project Diana.

Although my father never wavered in his gratitude to the Army, the gray-out of Project Diana, which he regarded as the signal scientific achievement of his life, always rankled. As he observed with just a faint hint of pique, "If we had been less vigorous, Dr. Zoltan Bay and his colleagues in Hungary would have been 'First' about a month later." He spent his last few years lobbying first for a full-scale 40-year commemoration and then, when that failed to materialize, for an even more lavish celebration of the 50-year anniversary. Had he lived he would have seen his hopes dashed once again.

Now it seems the tide may have turned, along with a growing realization that this story belongs not just to the Army but to the world. It has great visuals and a compelling narrative. Moreover, history has caught up with this achievement, and many of the preposterous predictions that once made the Army brass cringe have actually come to pass. Cell phones and global positioning systems have highlighted the importance of satellite communications. The popularity of Earth-Moon-Earth communication has increased recognition of Project Diana as the origin of this technology. NASA, with its genius for popularizing space-age technology and its general focus on peaceful applications of space exploration, has now become an important resource for information on Project Diana. Neil DeGrasse Tyson featured a contemporary newsreel clip in Episode 11 of his TV series *Cosmos*.

❋ ❋ ❋

Part of what was truly remarkable about this event, at least for me, was the venue itself. My father was actually being honored not only in his own country but in the very same World War II radar laboratory buildings where his seminal work was carried out.

The outside of these buildings was a very familiar part of the iconography of my childhood. On days when my mother anticipated needing the family car to run errands, she drove my father to work in the morning and picked him up again after work, my sister and me in tow. The evening ritual was always the same: We parked and waited, watching for a slight dark-haired man in a lumpy overcoat and fedora to emerge and take over the wheel. Never once did any of us go in, even my mother. At the time I'm sure I thought (correctly) that she didn't want to leave her little daughters alone in the car, but in retrospect, it also seems probable that these buildings contained top-secret materiel not meant for prying eyes and that no outsider, even a wife, was welcome.

Now, here we were inside—dining, sipping wine, and accepting accolades for his achievements.

All the credit for this unlikely scenario, and at least some of the credit for the current revival of interest in Project Diana, goes to the InfoAge Science History Learning Center and its tireless efforts to preserve the long and checkered past of the Camp Evans site. Just how unlikely? In 1993, with the end of the Cold War, the Department of Defense had decided to close many of its military bases, and Camp Evans was on the hit list. But for the remarkable vision of InfoAge founder Fred Carl, who steadfastly resisted the Army's demolition plans, the entire Camp Evans campus with all its rich heritage would have been bulldozed into oblivion. So—very unlikely indeed.

C. PROJECT DIANA

IN CELEBRATION OF DIANA DAY

On January 10, 1946, a small group of scientists at Camp Evans in Wall, NJ successfully bounced radio waves off the moon—and the world has never been the same. As Fred Carl succinctly put it, "Project Diana was a pivotal event that built on World War II expertise but pointed the way to the future."

In the 75 years that have elapsed since that day, we have

seen humans walk on the moon, we have seen the world transformed by the internet—and we have seen amateur radio enthusiasts moon-bouncing in their own back yards! We have seen the earth's population more than triple and witnessed a degree of globalization unimaginable in 1946.

A good moment to reflect on the larger significance of Project Diana.

Military Uses

In a very real sense, Project Diana was the opening salvo in the Cold War. Concerned that the Soviets had captured enough German expertise to build missiles capable of reaching the U.S., the Pentagon ordered Jack DeWitt, head of the Evans Lab, to develop radar systems capable of detecting and tracking such missiles. DeWitt interpreted his mandate broadly. Since there were in fact no such missiles in existence on which to perform such a test, he found in this directive the perfect excuse for pursuing his decades-old dream of "shooting the moon." "Well, if we could hit the moon with radar," he argued, "we could probably detect the rockets." With almost anyone else at the helm of Camp Evans, the project might have taken a different and probably less informative turn. Indeed, the successful moon shot appears to have taken DeWitt's superiors by surprise, and two weeks of checking and replicating elapsed before the War Department allowed the achievement to be announced to the world.

Although radar had proved highly effective in detecting enemy ships and aircraft during World War II, many doubted that radio waves could penetrate the ionosphere, bounce off a target, and be detected back on Earth. Project Diana served notice that anything the Soviet Union wanted to throw at us, including intercontinental ballistic missiles, could be detected and tracked as well. Development of weapons to intercept such missiles followed; the arms race and the quest for mutually assured destruction were underway.

Scientific Implications

Robert Buderi, in his book *The Invention that Changed the*

World, about the central role of the Massachusetts Institute of Technology Radiation Laboratory in the development of radar (for which the Project Diana constituted a bit of an inconvenient truth), dismisses the achievement as little more than a clever demonstration: "The hullabaloo soon died down. For all its remarkability as an engineering feat, the DeWitt experiment held next to nothing of scientific interest."

Is this really so? Project Diana was indeed a brilliant piece of engineering, but more than that, it disproved once and for all the going hypothesis that the ionosphere could not be penetrated by radio waves—and in so doing, lifted a veil in the field of astronomy. For most of human history, our knowledge of the universe was based solely on the visible portion of the electromagnetic spectrum. Adding the radio band to our toolbox extended our observational capabilities tenfold. As the first-ever experiment in radar astronomy, it offered the possibility not simply of passive observation but also of active probing of other celestial bodies.

Once the War Department conceded that Project Diana had indeed accomplished what Jack DeWitt claimed, it was quick to grasp the potential of moon bounce technology. Among its predictions: accurate topographical mapping of heavenly bodies, measurement and analysis of the ionosphere, and radio control of space travel, missiles, and orbiting artificial satellites. The news media also chimed in, with *Time* suggesting that Project Diana might provide a test of Einstein's theory of relativity, while a more skeptical *Newsweek* article labeled the War Department's predictions "worthy of Jules Verne." As a NASA historian concluded years later, "all of the predictions made by the War Department, including the relativity test, have come true in the manner of a Jules Verne novel."

Advances in Communication

Project Diana touches our lives every time we pull out our cell phones, the most ubiquitous example of satellite communications. Personal mobile communications devices have changed our lives and our expectations in ways ranging from how we deal with emergencies to the level of "sharing" we now expect from each other. (How many stories have you read and movies have you watched turning on plot twists that would not be possible had the protagonists had cell phones?) In many parts of the developing world cell phone technology is the open sesame to global participation. It does not replace landlines, it replaces *nothing*. It is all there ever was.

Project Diana is also the prototype for Earth-Moon-Earth communication, whereby amateur radio operators interact with other hams around the world by bouncing radio signals off the moon. My father and his team hoped to receive echoes from their own transmissions, whereas EME enthusiasts are more interested in sending out signals and having them received elsewhere on earth. But the technology is conceptually identical, and presumably so is the joy experienced by my father in being able to use the moon as a relay station for radio signals.

Helping to Shape the American Myth

Would the advances described above have happened without Project Diana? In some form, almost certainly yes. Several other labs were poised to mount similar demonstrations, and within a month of Project Diana's spectacular triumph, Zoltan Bay of Hungary, using an indirect detection method, also demonstrated that it was possible to reach the moon with radio pulses.

But Bay was second, and Project Diana was first, a fact that had a profound impact on the American psyche. Coming as it did on the heels of America's contribution to the defeat of the Axis powers in World War II, based on our "can do" spirit and technological expertise, it was so to speak the frosting on the cake, suggesting a host of peaceful as well as military applications for our capabilities. Since previous moon bounce efforts (including our own) had failed, and since many reputable scientists believed the ionosphere could not be penetrated by radio waves, Project Diana burnished our self-image as a people who could do the impossible. Because the team basically improvised, modifying equipment they already had, it reinforced our faith in our talent as engineers and tinkerers. Finally, it provided a cultural template for subsequent American space exploration, initiating the tradition of naming such projects after ancient Greek and Roman gods like Mercury and Apollo and glorifying (in the person of Jack DeWitt) the cowboy hero with the "right stuff" later exemplified by American astronauts.

Every country creates a mythology about itself that helps to form its national identity and make its people feel as though they are part of something larger than themselves. What aligns with the mythology is selected; what doesn't tends to be winnowed out. The postwar years were a time when America was starting to flex its muscles on a global scale, and its sense of itself was rapidly evolving to accommodate these developments. Project Diana was perfectly poised to be part of this process. For a variety of reasons, it is not so easy these days to feel exceptional on the global stage, which is probably why so many Americans harbor nostalgia for what in retrospect seem to them to have been simpler, better times.

THE CAST OF CHARACTERS

Mofenson, Webb, DeWitt, Stodola, Kauffman

What was so special about the men who somehow co-alesced into a team that was more than just the sum of its parts? What made them click? What allowed them to succeed where others had failed?

Key to the project's success was the visionary leadership of **John Hibbett "Jack" DeWitt, Jr.** A radioman from his teens, he founded station WSM in Nashville, Tennessee, famous for the

Grand Ol' Opry and for introducing country music to the world. When war broke out, DeWitt answered the call of duty and joined the Army. He was assigned to the Evans Signal Lab, the Army's hothouse for the development of radar, and by 1943 he had become the Director of the Lab, supervising thousands of employees. By the end of the war, over 1,000 remained on staff, mostly civilians like my father. The Army was preparing to re-tool for the anticipated Cold War, using less in-house labor and jobbing out a lot of its military research. So after years of working 6-day weeks, many of the radar specialists had little to do beyond wrapping up their unfinished business, and they were scheduled for decommissioning within a few months.

The plan to shoot the moon was probably hatched on V-J Day or not long after. It was a feat that was never far from DeWitt's mind. Far from being discouraged by the failure of an earlier attempt in 1940, he was all the more eager to do it again and get it right. When the Pentagon, concerned that the Russians had learned enough from captured German rockets to build missiles capable of delivering nuclear weapons to the U.S., assigned him the task of determining whether such missiles could be detected by radar, he quickly recognized that he'd been handed a second chance to pursue his dream. After all, there were no long-range missiles to practice on, Russian or other, so why not target the moon?

To my father, Jack DeWitt was both a friend and a hero. But perhaps most importantly, considering the circumstances that brought them together, he was a good boss: "Col. DeWitt was director of the Laboratory," my father wrote in a letter much later, "but he was no figurehead participant; he was the conceiver of the project and had undertaken trying to perform the experiment before he entered the Army—his original experiment failed because of known equipment inadequacies; he is a skilled engineer and his extensive participation was essential to the project's success."

To carry out the day-to-day work of correcting these "known equipment inadequacies" and assembling the complex array of components needed to do the job right, DeWitt decided to assemble a select group of highly experienced radar scientists to work with him. They needed to be responsive to his wishes but also able to think independently. They needed to have complementary but not totally interchangeable skills. They needed to be good problem-solvers who could come up with creative solutions on a limited budget. They needed to work well with him and with each other, not just on occasion but on a daily basis for several months.

They needed to have the mental flexibility to make a paradigm shift, from thinking of radar ranges in terms of a few hundred miles to a distance several orders of magnitude greater, around 240,000 miles. (Actually, the moon only barely tapped the capability of the Diana radar; calculations showed the maximum range was over a million miles.)

They needed to be men of discretion, able to maintain secrecy about the "Diana Project" until results were obtained and certified.

DeWitt was also aware that other scientists around the world were hoping to achieve the same goal. It was the sort of scenario familiar to science—stimulating each other's efforts by engaging in friendly or sometimes not-so-friendly competition to be The First. He knew he had to work quickly and efficiently—and quietly—if the Army team was to be the one to make history. He must have chosen his team of engineers with this consideration in mind as well.

And how did DeWitt decide just how many people he needed? According to Wharton management professor Katherine Klein, the widely-accepted optimal size for a working team is five people. More than that and individual performance may get lost, fewer and there may be skill gaps. A team of five is large enough to function as a social unit, but not so large it that breaks down into sub-teams. Although it seems improbable that DeWitt had studied modern team size theory, he seems to have hit intuitively upon the ideal formula for his purpose.

As Director of the Laboratory, DeWitt presumably had had

ample opportunity to observe the engineers who worked under his command. My father, as Chief of the 35-employee Special Developments Section of the Laboratory, probably also had input into staffing decisions. Jack Mofenson and Harold Webb were two of his assistants, and Herbert Kauffman rounded out the team. These five men were the ones, who, as Webb later wrote, [did] "most of the work on the Diana project," the ones who worked together in the lab day in and day out, often including weekends. Stodola, Mofenson, and Kauffman were radio engineers; Webb was a physicist. It was a young team. DeWitt, the oldest, turned 40 a month after the successful moon shot. Webb was three years his junior. Stodola, Mofenson, and Kauffman were all 31, all born within a few months of each other in 1914. Except for DeWitt himself, all were civilians.

My father's admiration for DeWitt must have been mutual, for of the 1,200 or so employees remaining at Camp Evans by 1946—about 70 officers, the rest civilians—he chose an elite team of five experienced radar engineers and of those, selected my father to serve as scientific director.

Edwin King Stodola was just the sort of man Jack DeWitt would like and respect—an engineer with good social skills and expertise in moving target detection (in this case a very large moving target). His approach to everything, from household repairs to defeating the kamikaze pilots to shooting the moon, reflected the can-do spirit that Americans brought to World War II —work rapidly and intensively, modify existing equipment, and then test, test, test.

My father was a born skeptic—"it ain't necessarily so" was his favorite song and his favorite saying. He was a stickler for correct grammar and a lover of words and their history—the longer and odder the better; not surprisingly he was an avid Gilbert & Sullivan fan. He was too restless for long intimate conversations —he couldn't even sit in a restaurant without excusing himself to make a phone call or two, leaving my shy mother alone at the table to fend for herself—but he was assiduous about nurturing his friendships and often drove miles out of his way for the briefest of visits. Everyone who knew him much later, when he took up flying, remembers his fly-in, fly-out visits often requiring the person being visited to meet him at an airport café.

Jacob "Jack" Mofenson and my father were already friends by the time they became Project Diana teammates—probably brought together shortly after Jack's arrival in April 1942 by the similarity of their life trajectories. Both men were born in Brooklyn and both received their engineering education at New York institutions legendary for merit-based admissions—my dad at Cooper Union, Jack at the City College of New York. Their oldest children—David Mofenson and I—were born less than two weeks apart and were each other's first playmates; both David and I had younger sisters born two years later. I have a home movie clip of several minutes of David and me taken a few months before we turned two, which in addition to an overdose of baby cuteness includes cameo appearances by the two sets of parents playing outside with us, both men sporting their fedoras.

Born in Johnson City, Indiana, **Harold Donovan Webb** was the only Midwesterner of the group. He was also the only one with a doctoral degree, from Indiana University, and the only academic; after leaving Camp Evans in 1947, he spent the rest of his career as a professor at the University of Illinois College of Engineering, where he studied the ionosphere, moon reflection, and radioastronomy. His daughter Diana remembers him as a quiet man who in later life chose not to play up his seminal role in the success of Project Diana—though at my father's request he tried unsuccessfully to persuade his cousin Jane Pauley to do a national news story on it.

Herbert Percival Kauffman was New Orleans born and bred; his mother was of Cajun descent. Once his wartime interlude on the Jersey Shore was over, he could hardly wait to return to his home and his radio career. His first-hand account of the moon shot—they all wrote at least one!—evinces a wry understated humor: the famous bedspring was "decidedly not a back-yard antenna" and the Project Diana team held "the new DX [long-distance] record."

In the aggregate, these men were a tribute to Jack DeWitt's team-building skills. They came to work every day with the strong conviction that together they could accomplish their goals, step by step. There was not a hotdogger among them, not one with a need to claim credit at the others' expense. When the *New York Times* announced the feat to the world on January 25, 1946, the reporter described the "five principals" as "modest in the extreme," having to be coaxed into revealing even a little biographical detail. Jack Mofenson was probably the most self-effacing: "I was a diamond dealer," he laughingly told the reporter. Yet they never lost their sense of awe; Herbert Kauffman described himself as "bowled over" when they actually succeeded. Harold Webb continued to ponder the moon bounce experiment in an effort to understand it better throughout his life.

Although they shouldered the bulk of the responsibility, DeWitt and his team were not working in a vacuum. "A lot of other people," as my dad put it, "got 'scrounged' into [the effort]"—often working off the clock. Participating mostly if not completely *in absentia* from his lab in Alpine, New Jersey was **Edwin Howard Armstrong,** the iconic radio pioneer who designed the receiver and transmitter modified for use in the moonshot. Input was also sought from representatives of other Evans Lab units in the form of scientific expertise, equipment, and hands-on know-how. Most notable of these was **Walter Samuel McAfee** of the Mathematical Analysis Section, who was responsible for the elegant calculations of the moon's location with respect to the earth on any given day. Also included on the list (gleaned from the Acknowledgments section of DeWitt and Stodola's 1949 paper in the *Proceedings of the IRE*) were **J. Ruze, O. C. Woodward, A. Kampinsky, J. Corwin, C. G. McMullen, W. S. Pike, F. Blackwell, R. Guthrie, H. Lisman, J. Snyder, G. Cantor,** and **A. Davis.** For some, being associated with Project Diana was a source of great pride; for the rest of his life, an assistant named Gilbert Cantor liked to boast that he had been King Stodola's "tool in hand."

Despite their mutual respect and close working relationship, the quintet of engineers went their separate ways after the completion of the project. But the bond that was forged during those long intense days remained, and they kept in touch for the rest of their lives through a friendly exchange of holiday cards and photos and the occasional visit or phone chat.

* * *

Postscript: Robert Buderi, in his book about the Massachusetts Institute of Technology Radiation Laboratory, was forced into a three-page detour to give an account of the moon shot, a feat the Rad Lab might have accomplished but didn't. In it, to the endless amusement of my family, Buderi refers to my father as "the diminutive, bespectacled E. King Stodola." Bespectacled, yes—but diminutive? Certainly none of us ever thought of him that way. Only recently did I realize that the photo of the five engineers shown above, which was widely published when Project Diana first hit the headlines in 1946, must have been the source of Buderi's description. My dad is the shortest, to be sure—but hey, someone's gotta be the shortest! Mystery solved.

JACK DEWITT: "A LUNAR LOVE AFFAIR"

Jack DeWitt adjusting transmitter controls

John Hibbett DeWitt, Jr. was many things to many people over the course of his nearly 93 years:

He was Nashville through and through, from his birth on February 20, 1906 to his death on January 22, 1999. His father was a judge in the Tennessee State Court of Appeals. He later attended Vanderbilt University and, after an interruption in 1929 to work in Bell Labs in Washington DC, in those days sort of a quasi-graduate school for aspiring engineers, he returned to complete his degree. For the rest of his life, he left Nashville only when he was needed elsewhere, and always returned when it became possible to do so.

He was a seminal figure in the Golden Age of Radio. In 1922, at the age of 16, he drew on his skills as an amateur radio operator (N4CBC) to launch Nashville's first broadcasting station in his parents' living room—the 15-watt WDAA, commissioned by the Ward-Belmont school for advertising purposes. WDAA lasted about a year.He then took on a much more ambitious project when he helped install the transmitter for WSM, Nashville's first commercial radio station, which took its name from the slogan of its sponsor, the National Life and Accident Insurance Company, "We Shield Millions." WSM first signed on in October of 1925 and a couple of months later introduced the show that would make Nashville and country music famous around the world, the Grand Ole Opry, now approaching its 95th year on the air. In 1932, after another stint at Bell Labs (by now in New York), he returned to Nashville to become chief engineer at WSM, where he oversaw installation of its new tower and 50 kW transmitter, operating on a clear channel of 650 kHz. After sundown under good conditions, WSM's AM signal can "skip" across the ionosphere, reaching perhaps 35 states. The cultural impact of bringing live country music and the diverse performers who made it into living rooms across America, if only for a weekly one-hour "barn dance", can hardly be overstated. From 1947 until his retirement in 1968, he served as President of WSM AM, FM, and TV. In an innovation that was pure DeWitt, WSM was the first TV station in the U.S. to broadcast near-real-time weather satellite photographs.

He was passionate about astronomy, an interest kindled during the three years he spent at Bell Labs between 1929 and 1932, and solidified upon his return to Nashville, when he found that his brother Ward had also been bitten by the astronomy bug and the two built a 12-inch Cassegrain telescope, grinding the mirror themselves. In 1947, back in Nashville once again after completing his mission at Camp Evans, he built a dry-ice refrigerated photoelectric photometer that subsequently served as the basis for many Vanderbilt masters' theses, bringing photoelectric photometry to Nashville and indeed to the whole Southern U.S. Although I and others think of him as first and foremost a radioman, the American Astronomical Society regards him as one of their own, stating in their obituary that "the world lost a pioneering astronomer."

At least temporarily, he was a military man—though perhaps less in the tradition of stoic and unquestioning obedience than in the Chuck Yeager tradition of having the "right stuff." When WWII broke out, he left WSM and his beloved Nashville to answer the call of duty and join the Army, becoming a pioneer in the development of radar when he was assigned to the Evans Signal Laboratory of the U.S. Army in Belmar NJ. By 1943, at the age of 37, he had become Director of the Signal Laboratory, supervising the work of many thousands of employees at the height of WWII.

And on January 10, 1946, he accomplished the goal for which he will always be best remembered, when the Project Diana team he assembled after the War ended successfully bounced radio waves off the moon.

* * *

The title of this essay, "A Lunar Love Affair," was borrowed from an overview of Project Diana by Trevor Clark that appeared in the *IEEE Spectrum* in May 1980. It all started when he became engrossed in the study of astronomy during his three years at Bell Labs. In 1935, back in Nashville, he attempted unsuccessfully to receive noise from the then newly-discovered Milky Way. Nothing daunted, he continued thinking about extraterrestrial communication, and by 1940, more specifically about the moon. On May 21, he famously wrote in his notebook: "It occurred to me that it might be possible to reflect ultrashort waves from the moon. If this could be done, it would open up wide possibilities for the study of the upper atmosphere. So far as I know, no one has ever sent waves off the Earth and measured their return through the entire atmosphere of the Earth." He went on to describe his attempt the previous evening to reflect 138 MHz (2.2-meter) radio waves off the moon, using an 80-watt transmitter and receiver he had developed and built for radio station WGN in Chicago. The experiment failed; his yearning for contact with the moon did not.

Fate handed DeWitt a second chance to attempt a moon bounce in 1945, when his demobilization was delayed for several months after World War II ended. He quickly assembled a small team of his most skilled engineers and gave them access to all of Camp Evans' resources, including in the Laboratory's Theoretical Studies Group, the Antenna and Mechanical Design Group, and others.

It was DeWitt who christened the experiment Project Diana, after the Roman goddess of the moon, stating a little crudely, "the Greek [sic] mythology books said that she had never been cracked." His choice led to the American tradition, which continues to this day, of naming space missions after figures from ancient mythology.

To Jack DeWitt, failures were just speed bumps on the road to success. He thought about his "negative result" in 1940 and carefully considered how his little team could tap available resources —time was of the essence and funds were limited—to maximize their chances of success. His to-do list included the choice of a transmitter that could generate signals capable of making the 480,000-mile round trip to the moon and a receiver that could amplify the returning signals to detectable levels, a decision about the optimal frequency of the signal needed to achieve these goals, the mathematical calculations needed to compensate for the Doppler shift from a moving object (the change in frequency between a transmitted pulse and a received echo), and the design of an antenna capable of transmitting and receiving these new signals. Under his guidance, all these tough problems were grappled with and successively overcome.

Initial efforts to hit the moon were frustrated repeatedly, by dodgy equipment and possibly (as later suggested by Harold Webb) by their ignorance of a phenomenon known as the Faraday effect that caused the signals to be rotated as they passed through the ionosphere and be diminished in a horizontally polarized receiving antenna.

Finally, on January 10, 1946, it all came together. The first signal was broadcast just before noon, and 2.5 seconds later—the amount of time required for a round trip of about 480,000 miles—the echoes lit up a 9" cathode-ray tube and produced a 180 Hz beep amplified by a loudspeaker. Minutes later, when the moon moved out of range, testing ended for the day. Although the source of the echoes was inferential, DeWitt later remarked that it had to be the moon "because there was nothing else there but the moon."

Ironically, DeWitt himself wasn't present on that fateful day. "I was over in Belmar," he confessed later, "having lunch and picking up some items like cigarettes at the drug store (stopped smoking 1952 thank God)." Testing was repeated daily for the next 3 days, however, and then on eight additional days during the month, so he eventually had ample opportunity to enjoy the fulfillment of his dream.

The equipment, however, remained "haywire," as DeWitt put it, and some of the media excitement that followed was actually based on simulations, recordings, and scripted interviews. When a couple of prominent colleagues from the Massachusetts Institute of Technology Radiation Laboratory arrived to observe a test carried out under my father's direction, what happened was…nothing. As DeWitt recounted the story later, "You can imagine that at this point I was dying. Shortly a big truck passed by on the road next to the equipment and immediately the echoes popped up. I will always believe that one of the crystals was not oscillating until it was shaken up or there was a loose connection which fixed itself." Cheers erupted from the bystanders.

Such a project clearly needed a military justification, and DeWitt found it in a directive from the Army's Chief Signal Officer to develop radars capable of detecting missiles coming from the Soviet Union. Since no such missiles were available for tests, DeWitt argued, the moon could serve as a handy stand-in. But DeWitt himself was more interested in its potential for space exploration, and radioman that he was, in the project's implications for communication. In his visionary notes of 1940, he wrote "There are times when communication by this method might be extremely valuable such as during magnetic storms and daytime radio 'blackouts'. This may provide a means in the future of bringing television programs over long distances, such as across the oceans." According to Harold Webb, Project Diana's potential for communication was still his obsession in 1945: "He thought TV signals could be bounced off the moon and spread to one-half the earth." Nary a word about its military implications was spoken, at least to those with whom he worked most closely.

✳ ✳ ✳

DeWitt shared with many of his fellow radio engineers, including Armstrong and of course my father, a fascination with towers—and the bigger the better.

In 1928, WSM was assigned the frequency of 650 kHz, giving it membership in the highly select club of Class 1-A clear-channel broadcasters—meaning that no other station in the country could share that frequency. To take advantage of all this power and obtain nationwide coverage, the station under DeWitt's supervision erected an unusual diamond-shaped vertical tower (manufactured by Blaw-Knox) in 1932 to support the station's new 50,000-watt transmitter. Topping out at 878 feet, it was at that time the tallest antenna in North America.

Can a tower possibly be *too* tall?

DeWitt and his crew of engineers soon noticed that the tower was actually causing self-cancellation in its "fringe" reception areas, keeping it from reaching fans in Chattanooga and Knox-ville, and in 1939 it was trimmed to a mere 808 feet. It is now known that the optimal height for a Class A station on that frequency is about 810 feet, so—close enough. The lopped-off portion was recycled as a flagpole at a nearby school, where it remained for more than 50 years.

During World War II, the tower was assigned to service as a backup relay station for transmissions to submarines should ship-to-shore communication be lost.

The tower, still standing proud just south of Nashville in Brentwood, has been designated a National Engineering Landmark and is listed in the National Register of Historic Places. In 2001, when the Country Music Hall of Fame and Museum moved to a new facility in the heart of downtown Nashville's arts and entertainment district, its design incorporated a replica of the distinctive diamond-shaped tower on top of the Rotunda, in recognition of WSM's revered place in country music history —thanks in no small part to its tower. It is among the oldest operating broadcast towers still in use, and for tower-philes every-where, a must-see.

HAROLD WEBB, SUPERHERO

S hortly before going to press, I received an email from my fellow Project Diana legacy, Diana Webb, along with photos of a 4-page story in the June, 1946 issue of Picture News starring—her father! "When I was a kid," she wrote, "I was proud to be able to tell my friends that my dad had been included in a comic book." What kid wouldn't have been? I knew I had to drop everything and write an essay on it.

❉ ❉ ❉

Comic strips had been around since the late 19th century, but comic books as we know them didn't appear until 1933—almost by accident. Printers had been casting around for ways to use equipment lying idle between jobs when someone came up with the bright idea of creating a comic section that could be folded down from a broadside into eight 9"x12" pages. From there it was just a hop, skip, and a jump to a freestanding booklet that could be sold separately.

Once the format was set, all that was needed for the genre to take off was a galvanizing event. That event came in June of 1938 with the introduction of Superman, marking the start of the "Golden Age" of comics.

Superman was soon joined in the epic struggle between Good and Evil by his fellow superheroes Batman, Wonder Woman, Green Lantern, and perhaps second only to Superman, Captain Marvel. Alongside the superheroes, other high-action genres gained in popularity, including science fiction, horror, crime, and stories of the wild, wild West. Once World War II began, Captain America, caped in the stars and stripes, also joined their ranks, integrating current events into the genre by supporting the war effort and doing battle with Hitler.

Starting in the early 1940s, however, the combination of un-relenting violence and unrealistic derring-do started to receive some pushback in the form of doubts as to whether all this was really a good thing for young, mostly male hearts and minds. A psychologist named Fredric Wertham, in a book titled *Seduction of the Innocent*, proclaimed that comic books "systematically poisoned the wellspring of children's spontaneity and prepared the ground for later aggressive behavior." In response to continuing pressure from Wertham and others, a self-regulating body called the Comics Code Authority issued a decree in 1945 stipulating that criminals must pay for their crimes and outlawing blood and gore, the use on the cover of words like WEIRD! and HORROR!—and depictions of two-piece bathing suits.

The combination of the Comics Code Authority and the postwar "return to normalcy" led to a falloff in sales and a decline of interest in superheroes. But never completely. Many of the classic comic characters managed to survive and still command an audience today. In addition, new forms arose to reframe the superhero—and it is in this category that I would place the portrait of Harold Webb in the short-lived *Picture News*, published by Lafayette Street Corporation. *Picture News*, presenting itself (not without some justice) as educational and harking back to the tradition of Captain America, drew its superheroes from current events, including sports, true-life adventure tales, and as in the case of Harold Webb, science.

Project Diana ticked all the boxes for this format. It offered an opportunity to explain the science behind Project Diana and to portray a real-life superhero in the person of Harold Webb. Best of all, it provided a jumping-off point for a discussion of its implications for space exploration by humans. As Fred Carl of InfoAge remarked, "The discovery that the ionosphere could be pierced and that communication was possible between earth and the universe beyond, opened the possibility of space exploration that previously had been only a dream in adventure films and comic books." The *Picture News* piece features a new breed of superhero—still with all the trappings of a comic book, but with the dream of interplanetary travel now anchored to a scientific finding.

The story includes two panels depicting Webb that, according to his daughter Diana, were drawn from photos that appeared in the news at the time, one gazing at an oscilloscope screen, saying "There it is!", the other talking about "Radio contact with rocket ships...." Diana suspects the latter speech was a paraphrase of what he actually said during the interview he must have given —it "does not sound like something that he actually would have said!" *Picture News* then takes a quantal leap to even wilder speculations of the kind that made the Army cringe—the title of the story, after all, is "A Round Trip to Mars!"

Diana also shared a photo of the cover of the issue because her father had signed it, but the picture itself has nothing to do with Project Diana. It goes with the *real* top story of the issue, about the upcoming rematch between prizefighters Joe Louis and Billy Conn at Yankee Stadium.

WALTER MCAFEE: THE MAN WHO DID THE MATH

Walter McAfee and King Stodola, reminiscing about Project Diana

A lthough Jack DeWitt's small team of engineers carried out most of the day-to-day work of designing and assembling all the parts that had to come together to succeed in their mission, they were also able to draw upon a lot of other resources available at Camp Evans for help and consultation.

Not all the "help" they received panned out. A novel antenna design proposed by two prominent specialists from the Antenna Design Section, for example, failed to deliver, and the engineers were thrown back on their own resources. Fortunately, the team came up with the bedspring design on its own, and the Mechanical Design Section succeeded in carrying out their instructions.

By contrast, one of their most valuable contributions came from the Mathematical Analysis Section, in the person of Walter Samuel McAfee (1914-1995), mathematician and theoretical physicist. McAfee was charged by DeWitt and his team with three tasks, all focused on ensuring that the outgoing signals would bounce back and then be detected by the receiver: 1) calculating the reflective characteristics of the moon; 2) calculating changes in the earth's position relative to that of the moon; and 3) working out the resulting Doppler frequency shift so that the receiver frequency could be correctly adjusted to detect incoming signals from the moon. The frequency of the incoming signal could vary as much as 300Hz from the outgoing signal. As DeWitt later put it, "We had to tune the receiver each time for a slightly different frequency from that sent out because of the Doppler shift due to the earth's rotation and the radial velocity of the moon at the time." Having tried and failed at an earlier moon bounce attempt in 1940, DeWitt was in a good position to appreciate the mathematical sophistication required to carry out these calculations.

The fact that McAfee was close at hand and available to the Project Diana team can be credited to the fact that early on, someone at the Signal Corps at Fort Monmouth was smart enough to recognize that African Americans represented an enormous untapped talent pool, and the post came to be known in the 1940s and 1950s as the Black Brain Center of the U.S. While it would be naive to romanticize Fort Monmouth as a utopia of nondiscrimination, it was known as a place where African

Americans had an unusual opportunity to become not just jani-
tors or technicians but scientists working at the highest level. As
one high official later remarked, it was "a place where, twenty
years before the civil rights movement, African Americans could
do great things"—this at a time when it was unusual for people
of different races even to be working alongside one another.

In 1942, after graduating from Ohio State University with an
M.S. in Physics in 1937 (his doctorate in Physics from Cornell
University came later, in 1949) and a stint as a junior high school
physics teacher in Columbus, Ohio, McAfee was recruited to fill
an opening for a civilian physicist at the Army Signal Corps. He
quit his teaching job with some trepidation, because contrary to
usual practice, the application form had not required a photograph and he wasn't sure what would happen when his new
employers discovered he was African American. What a relief to
find other desks already occupied by African Americans when he
arrived.

McAfee went on to have a distinguished career following his
work on Project Diana, including studies on radar coverage patterns taking into account diffraction around the curved surface
of the earth, the discovery that high-altitude nuclear explosions
can cause communications blackouts, and the development of
sensors used to detect and track enemy movements during the
Vietnam War. He remained at Camp Evans for 42 years, until his
retirement in 1985. In 2015 he was posthumously inducted into
U.S. Army Materiel Command's Hall of Fame, the first African
American to be so honored. Closer to home, the U.S. Post Office
building in Belmar was renamed for him in 1997. The McAfee
Center at Ft. Monmouth was named for him but later relocated
to Maryland when Ft. Monmouth was shuttered.

＊ ＊ ＊

The question of whether McAfee was given adequate recognition at the time for his role in Project Diana has been raised

in recent years, and there is evidence that McAfee himself felt he had been slighted based on his race. Unfortunately, though not surprisingly, there is no mention of McAfee's contribution, or even his name, in the media splash that directly followed the announcement of Project Diana's success. Only Dewitt and his team of four engineers, all of whom had worked at Camp Evans during the war developing military radar equipment, were introduced to the public at that time; and although McAfee belonged to one of several Camp Evans sections with whom the engineers consulted extensively and whose members might also have deserved a nod, it would have been—in view of the central importance of McAfee's contribution to the success of the project—not only the gracious thing to do but the right thing to do to acknowledge his work publicly at that time. It is a sad commentary that that did not happen, and in retrospect, an example of systemic racism.

I have not uncovered any evidence, however, that the engineers themselves withheld recognition from McAfee or downplayed his role in the scientific literature. Three of them wrote scientific papers shortly after the success of Project Diana: Jack Mofenson, writing in the April 1946 issue of *Electronics*, refers to "Calculations of the reflectivity coefficient made by Walter McAfee of the Theoretical Studies Group...." Harold Webb, in a report published in *Sky & Telescope*, also in April, wrote "Assuming that the surface of the moon is volcanic rock, non-conducting with a dielectric constant of 6, theoretical calculations made by Walter McAfee of Evans Signal Laboratories, show that...." Herbert Kauffman, in a piece published in *QST* in May, speaks of "Calculations made in the laboratory by W. McAfee and his Mathematical Analysis Section showed that...." All three papers then go on to describe the complexity of the challenge at some length, leaving no doubt as to either what the work involved or who did it. DeWitt and Stodola, taking the longer view in a paper published in *Proceedings of the I.R.E.* in March, 1949, also singled out McAfee for particular acknowledgment, citing his role in resolving "echoing-area problems."

I do know that my father and McAfee remained friendly and in touch. I love the charming photo posted above of the two men, both in their late 60s, reminiscing about the Diana days on the occasion of the 35th anniversary of the moon shot in 1981. Knowing my father's characteristic generosity in giving credit where credit is due regardless of race or gender, I am sure he would share my pleasure that McAfee's accomplishments, including his indispensable contribution to Project Diana, are now receiving the attention they deserve.

EDWIN HOWARD ARMSTRONG: DIANA'S GODFATHER

I n 1991, just a few months before he died, my father was awarded the Armstrong medal and plaque by the Radio Club of America. Of all the accolades he received, none would have been more meaningful to him. Sadly, he was suffering from Alzheimer's Disease and beyond grasping the nature of the honor that had been bestowed on him. He hadn't forgotten, however, that Major Edwin Howard Armstrong was one of his heroes, and he would happily discourse about the importance of Armstrong's work to anyone who would listen.

I now wish I had listened more closely.

* * *

Most radio buffs are familiar with Armstrong's turbulent career in wireless communications, during which he revolutionized the field not once but repeatedly—in the process stirring up mighty opposition from stakeholders in a new way of doing business that had little use for lone-wolf inventors like Armstrong.

His earliest work focused on improving receiver sensitivity. While still in college, he perfected the regenerative circuit, which dramatically improved radio reception by means of a positive feedback loop in the receiver, using a triode tube recently invented by Lee De Forest. Armstrong went on to invent the superheterodyne, which still further improved reception by mixing an incoming high-frequency signal with a second tunable lower-frequency signal to produce a predetermined intermediate frequency (IF) still further improved reception. The superheterodyne outperformed every previous approach including his own regenerative receiver and remains the industry standard to this day.

He then turned his attention to developing wide-band frequency modulation (FM) radio. Radios in use at the time were designed to be sensitive to the strength or amplitude of the incoming signal (that is, amplitude modulation or AM), but were also sensitive to environmental disturbances such as thunderstorms or electromagnetic waves emanating from electronic equipment. No amount of tweaking or shielding could fix this problem. Armstrong took a radically different approach, arguing that by varying the frequency instead of the amplitude of the signal to be transmitted and designing receivers accordingly, such interference could be prevented. He devoted much of the remainder of his life to demonstrating the superiority of FM.

Unfortunately for Armstrong, the commercial potential of the burgeoning field of wireless communication created a mercilessly competitive environment dominated by huge, well-heeled corporations. Armstrong's genius as a radio engineer was matched only by his naivete about the realities of organizational politics ("all substance and no style," as one biographer put it). Wildly underestimating the ability of greed and self-interest to prevail against (as he saw it) simple truth and honesty, Armstrong engaged in a long series of time-consuming, expensive, and sometimes quixotic legal battles to defend and protect his own interests.

His first such encounter was with Lee de Forest, who responded to Armstrong's success by laying claim to the idea of regeneration, despite little evidence that he even understood how the triode tube he invented worked, let alone that he had dreamed up Armstrong's brilliant new application for it. The ensuing litigation lasted for over a decade, with AT&T throwing its muscle behind de Forest after buying up his patents. In the end the Supreme Court, befuddled by the technical details, ruled against Armstrong, despite universal recognition among his scientific peers that regeneration was his invention and not de Forest's.

His conflict with de Forest, personally and professionally devastating though it was, paled in comparison to that subsequently elicited by the introduction of FM technology. By essentially eliminating the static that bedeviled AM radio, FM threatened the broadcasting industry not only by obsolescing millions of dollars worth of existing radio equipment overnight but also by diverting interest, attention, and coveted frequencies away from the anticipated Next Big Thing, television.

The Radio Corporation of America (RCA)—led by David Sarnoff, formerly his friend and collaborator, now his bitter foe —was not about to take this lying down. Both fair means and foul were employed to thwart Armstrong: Lawsuits were filed and then intentionally dragged out, patents were infringed, royalties were withheld, reverse engineering was used to buttress fake claims of priority. Armstrong was forced to remove his equipment from the top of the Empire State Building, ostensibly to make room for television equipment, driving him to move his operation to Alpine, New Jersey. Here the first FM station, W2XMN, began broadcasting in 1939—but only after the FCC first revoked his license and then restored it but diverted FM into a new frequency band at limited power—again, supposedly to make way for TV channel 1. (Ironically the Alpine station was briefly resuscitated after radio communication from the World Trade Center came to an abrupt halt on 9/11.)

Faced with the prospect of seemingly unending legal battles he could ill afford, Armstrong became despondent and even lashed out at his beloved wife Marion, who moved out of their home to escape further abuse. On the night of January 31, 1954, Edwin Armstrong donned his overcoat, scarf, gloves, and hat, removed the air conditioner from a window of his 13th floor apartment in Manhattan's exclusive River House, and jumped to his death. Marion Armstrong continued to prosecute her husband's unresolved infringement suits and ultimately triumphed, winning some $10 million in damages. Sadly, this vindication came too late to comfort or benefit Armstrong himself.

✳ ✳ ✳

Less known—perhaps because it remained highly classified for many years—is the story of Armstrong's work on FM radar during World War II. Indeed, in Armstrong's entry in the *Dictionary of American Biography*, this phase of his career is casually dismissed as "a hiatus caused by World War II."

Politically conservative and proud of his military service in World War I, Armstrong generously extended to the U.S. government royalty-free use of the patents he had so fiercely defended —a patriotic but costly gift. Meanwhile, in addition to his legal expenses, he was self-funding much of his research at Columbia (having declined a salary for his appointment as a full professor at Columbia in order to escape administrative duties and minimize teaching responsibilities), as well as his high-powered FM station in Alpine. (His red and white antenna, all 425 feet of it, still looms over the surrounding Palisades landscape, where its affluent neighbors regard it as an eyesore.)

As his debts mounted catastrophically, his attorney, Alfred McCormack, urged him to accept government contracts for his investigations of long range radar. These contracts enabled Armstrong to hire an assistant, Robert Hull, a newly-minted Columbia graduate, and together the two set about adapting FM technology to radar. The end of World War II, however, brought these explorations to a close, leaving no clear indication of what they hoped to accomplish. Since then, continuous-wave FM radar has found only specialized applications, and pulse radar remains the technology of choice for most purposes.

* * *

And here begins my quest to clarify the nature of Armstrong's role in Project Diana. Starting with a standard SCR-271 early warning radar, Armstrong and Hull, realizing that higher average power, much longer transmitter pulses, and a very narrow receiver bandwidth would be required, transformed the equipment into a powerful transmitter and sensitive receiver. The Project Diana team modified the set still further, disabling the FM modulator and adding a tunable crystal to allow the narrowband receiver frequency to be adjusted to compensate for the Doppler frequency shift caused by changes in the earth's position relative to that of the moon.

As I progressed with my research on Project Diana, I found myself becoming increasingly curious about whether Armstrong had directly interacted with the Project Diana team, and whether he had actually spent time with them at Camp Evans during this period. On the one hand, I had never, among all the first-person accounts I'd read by the Project Diana team, including my own oral history interview with my father, encountered any mention of face-to-face meetings or discussions with Armstrong. On the other hand, Belmar and Alpine are less than 100 miles apart, and Armstrong was highly familiar with the Marconi facility, where he and David Sarnoff in happier times had first listened to signals from his regenerative receiver.

The answers to these questions proved surprisingly elusive, even after I consulted such authoritative and comprehensive sources as *Empire of the Air* by Tom Lewis, *Man of High Fidelity* by Lawrence Lessing, *The Invention that Changed the World* by Robert Buderi, and the librarians in charge of the Armstrong archives at Columbia University. Finally, with the help of Fred

Carl, Director of the InfoAge Science History Learning Center, I found my way to Al Klase and Ray Chase, historians at the Radio Technology Museum at InfoAge, specializing respectively in the accomplishments of Edwin Howard Armstrong and the development of radar. Al was kind enough to direct me to an audio recording made in 2005 of an interview with Renville McMann, a panelist at a celebration of the 70th anniversary of FM radio, broadcast from Alpine. In this loving reminiscence of Armstrong, McMann describes the time he innocently suggested that Armstrong point his equipment towards the moon—and with uncharacteristic vehemence, Armstrong refused. That feat, as McMann later learned, was reserved for the Army. "Armstrong had a duplicate setup of the Camp Evans equipment at Alpine," adds Al; indeed, "the SCR-271 radar tower, sans antenna, is still there…. So clearly, there was direct contact with the Diana team. Armstrong's narrow-band receiver was crucial to the success of the project."

Al goes on to observe, "It's easy to assume Armstrong visited Camp Evans during the Diana era, it was only a day trip, even without modern roads, but I see no hard evidence. Dave Ossman, in his excellent radio drama version of *Empire of the Air*, has Armstrong at Evans for the first experiment, but rereading [the original book version of *Empire of the Air* by Tom Lewis, p. 298], we could attribute this to artistic license. I, too, would like to know if he was there."

But as Ray muses, it appears that the famous radio pioneer took pains to maintain an "arm's length" relationship with the youthful Diana team to ensure that they got full credit for whatever successes they achieved. He did such a good job of covering his tracks that barring some unexpected scholarly find, the nature and extent of his personal interactions with the Project Diana team will remain shrouded in mystery.

THE FAMOUS BEDSPRING ANTENNA

The bedspring antenna seen from below

This sweet anecdote was recounted by my cousin Tricia, who was twelve years my senior and stayed with my family off and on at several points during my childhood:

One evening King came home beaming and hugged [my mother] Elsa, saying "We did it!" He wouldn't tell me what [it was], but they were both ecstatic. Your mother's comment was a matter-of-fact "I knew you would" It was, I later learned, that they had bounced radar off the moon."

Not everyone, however, had shared my mother's implicit faith that Project Diana would succeed. Radar had more than demonstrated its utility during World War II in locating enemy aircraft and submarines, but—hit the moon? Earlier unsuccessful efforts had convinced many that asking radio waves to pierce the ionosphere, hit a designated object in space, and then return back through earth's atmosphere to the point of origin was expecting too much of the technology. Indeed, the standard method for measuring the distance to the ionosphere at that time was "pulse ranging"—that is, bouncing radio waves **off** the reflective surface of the ionosphere and timing their return, not passing **through** it.

Only a group of optimistic visionaries would attempt such a feat. Only a group of engineers would have the practical know-how to accomplish it.

*   *   *

If ever there was a monument to American ingenuity, it is surely Project Diana. Because the project was short on both time and money, the approach was to adapt materials already on hand. No attempt was made to design any of the main components for the project from scratch, and little if anything new was purchased to make it happen. The transmitter, the receiver, and the antenna all represented novel applications and redesigns of equipment they had used before.

These three elements were interdependent and had to work together as one for the project to succeed. Arguably the most critical, however, was the antenna, since the failure of previous attempts was attributable in large measure to insufficient sensitivity of the receiver antenna.

* * *

By the end of World War I, Army scientists at Fort Monmouth realized that the biggest threat in the next war would come from the air, and that Americans could no longer depend on the Atlantic and Pacific Oceans to protect and isolate them. Early detection of incoming threats was crucial if attacks on the homeland were to be avoided or minimized. Initial conceptual explorations of radar began as early as 1920. In the mid-1930s, radar research at Fort Monmouth took a more practical turn, to the point where the development of a prototype of the SCR [Signal Corps Radio OR Set Complete Radio, used interchangeably]-268 was well underway.

Then, in 1938, shaken by the discovery in their midst of a Nazi spy named Guenther Gustave Maria Rumrich, a Chicago-born German whose friendly curiosity had enabled him to infiltrate the radar research program alarmingly easily, the Army decided to increase security and reduce accessibility by moving the operation from Fort Monmouth to Fort Hancock.

It was at Fort Hancock that the legendary SCR-270/271, capable of detecting bombers 150 miles away, was developed under the leadership of Dr. Harold Zahl. Among other important innovations, this system featured a common antenna for both transmitting and receiving, made possible by a gas-discharge device called a duplexer invented by Zahl. The SCR-270 was a mobile unit; the SCR-271 was a fixed, tower-mounted version that differed mainly in having an antenna with a somewhat higher resolution.

My father was recruited as a young engineer to the Army's radar research program at Fort Hancock in early 1941, in the Radio Position Finding section. He participated in the development of the SCR-270/271 radar and continued to work on modifying and improving it throughout the war. Many years later, he wrote of this radar that it was "still the Old Faithful, coming through where more modern and more advertised sets have become unavailable."

As the nation inexorably drifted towards war, it was realized that Fort Hancock, at the tip of a peninsula surrounded by the Atlantic Ocean, New York Bay, and Sandy Hook Bay, also had some shortcomings as a location for radar research. More space was needed, it was argued, and the fierce nor'easters that periodically struck the base coated the radar equipment with a film of salt that undermined performance. Its location, so favorable to defense from conventional land and sea attacks, made it vulnerable to U-boat strikes.

So the Army purchased the old Marconi site in Belmar from King's College and rechristened it Camp Evans. Piecemeal, the radar research program was evacuated from Fort Hancock. Pearl Harbor hastened the transfer, and by 1942 the move to Camp Evans was complete.

With the acquisition of the Camp Evans site, the Signal Corps inherited a rich history of antenna development. During World War I, the Navy had assumed control of the property, and although Marconi's famous 400-foot wireless towers were used for the dispatch of important messages—indeed, historians dubbed World War I the "wireless war"—breakthroughs in reducing radio static achieved by a resident Canadian scientist named Roy Weagant enabled replacement of these ungainly structures by safer and cheaper if more ho-hum 30-foot antennas. This news was kept under wraps until after the war; "The End of the Giant Towers," proclaimed contemporary headlines. All of them were gone by 1925.

* * *

The arrival of the Army and the entry of the U.S. into

World War II transformed the site into a major center for radar research. In addition to newer systems (some involving testing of conceptual designs developed at the Massachusetts Institute of Technology Radiation Laboratory), refinement and stepped-up production of the "Old Faithful" SCR-270/271 continued throughout the war—turning Building 37 into a veritable antenna factory as bedspring-type array antennas ranging in length from 2 to 30 feet were assembled for installation on trailers to be used in remote locations. One such SCR-270 radar provided an early warning of incoming Japanese aircraft at Pearl Harbor—though in one of history's most egregious command and control failures, the information was initially misinterpreted by the Operator and subsequently discounted by the Commanding Officer on duty.

Project Diana, a code name intended (among other things) to befuddle the Guenther Rumrichs of the world, introduced a whole new series of challenges beyond the reach of existing technology designed for (relatively) short-distance detection of enemy aircraft.

Early on, Jack DeWitt decided to use as a starting point the crystal-controlled FM transmitter/receiver specially designed for the Signal Corps by radio pioneer Edwin Howard Armstrong. Armstrong's transmitter was in turn based on the SCR-271, which could operate at a wavelength of 2.7 meters—short enough, DeWitt and his team believed, to penetrate the ionosphere.

The Armstrong set had features that could be adapted to address two important problems:

First, due to the relative velocities of the earth and moon, the frequency of the returning echo differed from the transmitted signal (a phenomenon known as Doppler shift) by as much as 300 Hz, a number that was constantly changing depending on the earth's rotation and the moon's orbital path. DeWitt called on the Theoretical Studies Group for these calculations, which were carried out by the gifted mathematician Walter McAfee. The modified Armstrong radio was capable of being fine-tuned to the exact frequency required to compensate for the Doppler shift at a given point in time.

Second, signals bounced off an object 240,000 miles from earth would take much longer to echo and be much too weak to be detected by receiving antennas then in use, a problem that had bedeviled previous attempts to shoot the moon. Several steps were taken to enhance the detectability of the echo. "We realized that the moon echoes would be very weak," DeWitt later recalled, "so we had to use a very narrow receiver bandwidth to reduce thermal noise to tolerable levels." In addition, the power of the transmitted signal was increased. To further augment the returning signal, it was decided to generate a much longer pulse that would be easier to detect. The Armstrong radio was one of the few existing sets capable of generating such a long pulse.

To ensure successful reception of the echo, however, one more major problem remained to be resolved, that of antenna sensitivity. There was no antenna on-site that was up to the job. To come up with a solution, DeWitt called on two very senior specialists from the Antenna Design Section, one the Section head, who proposed a clever system using quarter-wave step-up transformers. The only problem: It didn't work, even after extensive efforts to tweak the transmitter.

DeWitt then turned to his own little group, which came up with the inspired idea of positioning two SCR-271 stationary radars side-by-side to create an enormous (40x40-foot) double bedspring antenna consisting of an 8x8 array of 64 half-wave dipoles with reflectors that further enhanced the 111.5 MHz signals. Herbert Kauffman, writing in 1946, credits King Stodola and two of his assistants, Harold Webb and Jack Mofenson, with developing this approach. Translating that idea into reality was undoubtedly easier said than done, but the Mechanical Design Section was able to implement this alternative plan and assemble the so-called "bedspring antenna" that eventually became the lynchpin of the operation. The resulting array was 152 times (~22 decibels) more sensitive than a single dipole. Unfortunately, engineering specifications were destroyed by the Army in 1971, so our knowledge of the exact design details remains sketchy.

This whole contraption—there is no other word for it—was mounted atop a 100-foot reinforced tower in the northeast corner of Camp Evans. The heavy and ungainly antenna could not be tilted, it could only be rotated in azimuth; so moonshots could only be attempted twice a day, usually at moonrise but occasionally at moonset, during the 40-minute window that opened when the moon passed through the 15-degrees-wide beam of the antenna pattern.

The iconic Diana Bedspring, perhaps the most famous antenna in history, has become the unofficial symbol of Project Diana. Its picture, doctored by a photo editor who thought the moon looked too dim in the original so used the contemporary equivalent of Photoshop to "burn in" a picture of the sun in its place, appeared on the front pages of newspapers and on magazine covers throughout the world. It appears on the cover of this book. If the late David Mofensen's dream of having Project Diana commemorated on a postage stamp is ever fulfilled (which can't happen before 2046 because—I know, I checked—the U.S. Postal Service will only issue such stamps in multiples of fifty years after the event), surely the Diana Bedspring will be the featured image.

✻ ✻ ✻

So where is the Diana Bedspring? Perhaps it was scrapped in the mid-1950s, when the Army removed its old radar units from the site to make way for a parabolic dish antenna, a design already in widespread use by most U.S. search radar at the end of the War. Or perhaps it was finally destroyed in the early 1990s, when the Army began to demolish the historic site in response to a Department of Defense decision to close many military bases including Camp Evans—until Fred Carl, founder of the InfoAge Science History Center, almost literally threw his body in the path of the wrecking ball. Or perhaps it is still being stored in pieces in some forgotten location, awaiting reassembly. In any event, it is, in the wistful words of an InfoAge volunteer, "no longer available."

The replacement antenna was a 50-foot dish created by the Signal Corps, using the frame of a captured Nazi Wurzburg Reise radar, to serve as our first satellite tracking antenna. In deference to its predecessor, it was dubbed the "Diana Dish." In 1957, when the Soviets stole a march on the U.S. by launching their Sputnik satellite, the Diana Dish was joined by a companion 60-foot dish named the "Space Sentry." In 1960, control of space research was transferred to a new civilian agency, NASA, which continued the weather observation research already underway at the Diana site and proudly broadcast the first televised images of cloud movements from space.

The Diana Dish has joined the Diana Bedspring in the dust-bin of history, but the Space Sentry was gifted by the Army to InfoAge, which has refurbished it for scientific and educational purposes as part of a larger effort to restore and preserve the Diana site. On January 10, 2016, and again on January 10, 2021, the 70th and 75th anniversaries respectively of the first successful moon bounce, the Space Sentry has been pressed into service to make a series of Earth-Moon-Earth contacts.

Although the development of communications satellites has obsolesced moon bounce as a national security tool, Earth-Moon-Earth or EME has become quite popular among amateur radio operators. The object of modern EME, unlike Project Diana but like successor projects such as PAMOR, is a two-way exchange of information, in which a signal is sent from one station to another by ricocheting it off the moon; overcoming the challenges of weak signal communication is part of the fun and represents a test of skill. EME is the longest path between any two stations on earth.

To accomplish this feat, EME enthusiasts need to erect antennas that by amateur radio standards are often large enough to jangle the nerves of Property Owners' Associations and perhaps violate the aesthetic sensibilities of their spouses. By Project Diana standards, however, most modern EME antennas are mere minikins and some are even portable.

NOT THE FIRST?

outside experts (far left, far right) called in to confirm the moon bounce

Jonas Salk, developer of the polio vaccine, once made a wry observation about the stages through with a new idea or discovery passes before it eventually becomes an accepted truth: "First it is said that 'It can't be true'; then 'If true, it is not very important'; and finally, 'We knew it all along'."

Case in point: Project Diana.

Stage 1: It can't be true.

The successful moon shot was made on January 10, 1946 but wasn't announced to the world until January 25. Of course in those days events weren't tweeted and retweeted at warp speed almost before they are finished happening, but still, why the 15-day delay? At least part of the explanation was offered by my father when I interviewed him in 1979, describing the immediate aftermath of the event:

"And to make a long story short, the thing came off successfully. Then we let our bosses know what was going on—and they didn't believe it! They didn't believe we'd really done it. So they called in a number of outside experts…. One of them was a fellow named Waldemar Kaempffert, who was then Science Editor of the *New York Times*. He was a pompous fellow. He came down and we talked with him and very quickly convinced him that we were really getting echoes from the moon. And then another man that I also kept quite friendly with, Donald Fink —he was then Editor of *Electronics*….—[He] came down, and he was smart, he knew right away that there was no question about what we were doing."

Stage 2: If true, it is not very important.

After the news was announced, Dr. Harlow Shapley, Director of the Harvard College Observatory, was asked by a reporter to provide expert commentary. Dr. Shapley termed the feat "an interesting tool in exploring the solar system" but added that astronomers were working on other wartime discoveries that he predicted would be "far more startling than the radar contact when they were announced."

Unsubstantiated intimations of unspecified breakthroughs as yet unrevealed—thank you for that, Dr. Shapley.

Stage 3: We knew it all along.

Of course, no feat like Project Diana ever occurs in a vacuum. The war had spurred radar research in labs all over the world, on both sides of the conflict, and several other teams were poised to carry out a similar demonstration. Indeed, the great Hungarian scientist Zoltan Bay succeeded in his own attempt to reflect radio pulses off the moon less than a month after Project Diana using an indirect, electrolytic detector, an accomplishment for which he richly deserves recognition and respect. It is an injustice to imply, however, as do some historians, that it was a colossal stroke of bad luck that Bay didn't get there first, completely discounting the thousands of hours of work that went into making Project Diana a success—not to mention the fact that the driving force of this effort was Jack Dewitt's "lunar love affair," as someone put it, dating back to the 1930s. Project Diana was hardly a fluke!

Another team that perhaps could have gotten there first (but didn't) was the Massachusetts Institute of Technology Radiation Laboratory—compelling Robert Buderi in his book *The Invention that Changed the World*, about radar but mostly about the Rad Lab, to devote three pages to what he called this "Army coup"— because clearly no history of radar would be complete without Project Diana.

In the snarkiest twist of all, a comment was posted on the InfoAge Camp Evans website stating, "In fact, the Diana project was not first. Germans detected Moon radar echo in 1943 when developing the Mammut long-range radar, at ~600 MHz. They had to keep it secret, and only in ~1950 the engineer who saw Moon echo, earned his Dipl.Ing. degree for his work."

A German engineer and not the Project Diana team deserves the credit for being the first to bounce radio waves off the moon? Really?

A different perspective on this earlier moon contact is offered by Magnus Lindgren in his master's thesis submitted in 2010 to the Chalmers University of Technology in Goteborg, Sweden, in which he notes that "operators of a German experimental radar succeeded in hearing their own lunar echoes in January 1944, by pure chance."

There is a big difference, as Lindgren goes on to point out, between accidentally hitting the moon because it happened to get in the way when you were really trying to do something else vs making a *"deliberate"* (Lindgren's word), carefully planned effort to hit the moon based on antenna design and location, elaborate calculations of the moon's location, etc, thus "determining with certainty that radio waves could penetrate the Earth's ionosphere. This discovery," Lindgren continues, "was a prerequisite for all space-related communication projects to come, thus marking the beginning of the space age."

* * *

Primacy is a funny thing. A lot of scientists have failed to achieve it because they asked the wrong question, or because they asked the right question at the wrong time and the technology was not there to support it, or because the vision needed to keep their objective in focus was lacking. Yes, there is always an element of luck involved, but planning, timing, and a perfect match between the goal and the resources available to achieve it are also required.

On January 10, 1946, all these criteria were met. The Project Diana team aimed their radar at the moon, hit it, and recorded the echo back on earth a little over two seconds later. Then they did it again, and again. They were the first to do this. Ever. End of story.

A CULTURE OF SECRECY

After months of intense activity and preparation, my father and his colleagues had "shot the moon" with radio waves, and the moon had echoed back: message received.

So success had crowned their efforts, and the tiny band of men at the Evans Signal Laboratory had scooped much bigger labs at home and abroad. The news that our horizons were no longer limited by the earth's atmosphere, and that communication with extraterrestrial bodies was now possible, made front-page headlines all over the world, followed shortly by newsreels shown before the Saturday matinee and extended radio interviews.

And the rest is history, right?

Not exactly.

In fact, although the successful moon bounce occurred on January 10, 1946, the Army actually sat on this news for more than two weeks before finally revealing it to the world on January 25. Some of the delay was undoubtedly dictated by a desire to have outside experts review the team's claims, but an abundance of caution cannot entirely account for the aura of reticence surrounding the event. Very little information was initially revealed about even the five principals, who were described in the first *New York Times* article about the feat as "modest in the extreme. Only at the reporters' insistence was there any revelation of biographical material on the quintet." According to the late David Mofenson, son of team member Jack Mofenson, "My father used to like to quote Lincoln's Gettysburg Address: 'the world will little note nor long remember....'"

So why was the Army so reluctant to exercise its bragging rights? Perhaps its pride in Project Diana was tempered by a lingering concern that too much had been revealed about their capabilities (which just a few months earlier might have resulted in Camp Evans being a target for German or Japanese bombs). Or possibly Jack DeWitt, though not actively deceptive, had not been as forthcoming about what he was up to as the Army higher-ups might have wished, as per the last sentence in the following passage from the oral history interview I conducted with my father in 1979:

> "[Jack DeWitt] had thought for a long time about the possibility of getting enough energy onto the moon to be detectable back on the earth, and with this cadre of people that I was heading and the various pieces of apparatus that were around, we did some thinking about it and decided we had the resources to do the trick, so... we started to work on this moon radar project, and a lot of other people got "scrounged" into it. And to make a long story short, the thing came off successfully. **Then we let our bosses know what was going on....**"

My father, who justifiably regarded Project Diana as his life's proudest achievement, solicited support from friends with media connections in the vain hope of achieving national recognition of the event on the 35th anniversary in 1981 and again on the 40th anniversary in 1986. David Mofenson tried to get the U.S. Post Office to issue a stamp commemorating the 50-year anniversary in 1996: "I wrote our U.S. Senators etc but could generate no interest—more important to put out stamps celebrating Daffy Duck, Bugs Bunny!" Previous stamps have featured the lunar landing, and David might have been pleased to know that on February 22, 2016, the USPS issued a $1.20 Forever stamp showing a beautiful photograph of the full moon. So far no mention of Project Diana, though.

The Army's reluctance to take credit for Project Diana didn't end there. Decades later my father, then in a high-level position in Electronic Warfare at the Pentagon, was unable to gain access to records of his own work on Project Diana—on the grounds that he lacked sufficient security clearance! Even allowing for the stricter rules on classification and declassification that prevailed in those days, this denial seems unduly punitive.

For ham radio operators and Earth-Moon-Earth enthusiasts, and for historians of space exploration, Project Diana is a big deal. The InfoAge Science History Museum, which has been heroic in staving off the Army's efforts to dismantle the Project Diana site, hosted a gala event on the 70th anniversary of Diana Day in 2016, in which museum volunteers, the Ocean Monmouth Amateur Radio Club (OMARC), and Princeton University, reenacted this historic milestone. So far as I know, however, the Army has remained discreetly on the sidelines, posting occasional sedate press releases but leaving others to commemorate the event with the acclaim it deserves.

THE MOON ENTERS
THE COLD WAR

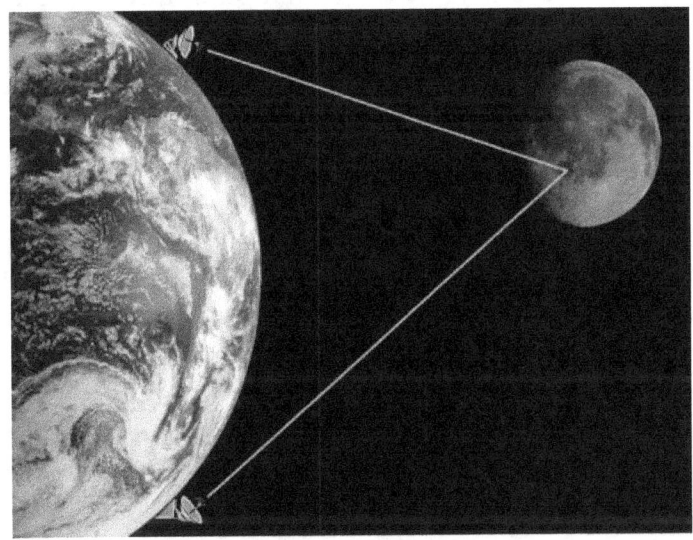

moon bounce via bistatic transmission (Courtesy of Roger Shultz)

P earl Harbor marked a shift away from American isolationism—not just as a political ideology but also as the
comfortable assumption that its location beyond two
oceans could somehow protect the U.S. from the rest of the
world. Pearl Harbor, and the declarations of war by Hitler and
Mussolini that followed shortly thereafter, thrust the U.S. into
the role of defender of liberty and democracy and leader of what

later came to be called the "Free World."

After the war ended with the Axis powers soundly defeated, the temporary alliance between the Western bloc, led by the U.S., and the Eastern bloc, led by the Soviet Union, unraveled due to profound social, economic, and political differences. The result was a Cold War between these two superpowers that lasted for over forty years, sustained by the belief on both sides that only the buildup of arsenals capable of "mutually assured destruction" could keep either side from demolishing the other. It was also characterized by the development of spy technology far more advanced than anything that preceded it—technology that, to remain effective, demanded almost epic levels of secrecy by those in the know.

I have often referred to Project Diana as the opening salvo in the Cold War, but only recently have I come to appreciate the full truth of this statement. Although much has been made of Jack DeWitt's almost obsessive fascination with the idea of bouncing radio waves off the moon, it is well to remember that his actual assignment from the Pentagon was to study ways to detect and track Soviet rockets that might, courtesy of expertise captured from the Germans, be capable of reaching the U.S. DeWitt argued, with some justice, that even though no such rockets were yet available for testing, hitting the moon would confirm that radar could detect them no matter how high they flew.

Not long ago, I learned of a recently declassified document, originally published in 1967, in which the author, Frank Eliot, asserts that the "entirely new technique" emerging directly from Project Diana—that is, using the moon to receive and reflect radio signals—offered a possible solution to the thorny problem of how to intercept Russian radar signals in an era when flights over the Soviet Union were prohibited. Although Project Diana involved *monostatic* transmission—that is, sending and receiving signals in the same location—a Naval Research Laboratory engineer named James Trexler figured out as early as 1948 that signals emanating from one location (e.g., in Russia) could potentially be detected via *bistatic* transmission to other locations (e.g., in the U.S.) if they happened to bounce off the moon.

Thus was born the highly classified P̲Assive M̲O̲on R̲elay or PAMOR, code-named "Joe," developed to spy on Soviet military radar signals reflected off the moon. Initial tests proved so promising that the project was intensified, at even deeper levels of secrecy. As one wag put it, "Leave it to the U.S. Navy to weaponize the moon."

The Russians, of course, weren't bouncing signals off the moon for the benefit of their Cold War adversaries. Indeed, most of their signals simply escaped the atmosphere and disappeared into outer space. Detecting those that did serendipitously hit the moon could only be done at certain times of day, in certain locations on earth where both the sending and receiving elements could "see" the moon at the same time; nor could the lunar terrain be too "rough" for clear reflection of signals. Antennas had to be at least 150 ft in diameter and preferably larger, and only a few were available for that purpose (e.g., those at the Grand Bahama tracking station, the Naval Research Laboratory's Chesapeake Bay Annex, Stanford University, and Sugar Grove, West Virginia).

There were more ways for this effort to fail than to succeed, but the military got lucky—not only in having favorable antenna locations and encountering favorable lunar conditions, but also, in the case of the "Hen House", a major anti-ballistic missile operation deep within the Soviet Union, in being able to take advantage of occasional brief practice sessions during which the Russians actually set their radar to track the moon. In the end, PAMOR proved to be an intelligence coup, continuing to yield information until the late sixties, when it was obsolesced by communications satellites.

PAMOR's success led the Naval Research Laboratory, in the mid-1950s, to commission an ambitious spinoff code-named Operation Moon Bounce, a series of experiments to test the feasibility of using the moon as a natural communications satellite. These tests were so effective in refining moon bounce technology that Operation Moon Bounce was used for several years to link Hawaii with Washington, DC. Like PAMOR, Operation Moon Bounce was superseded in the late 1960s by networks of communications satellites—networks whose design profited from the experience gained during the Moon Bounce tests.

Moon bounce communication, generally referred to as Earth-Moon-Earth or EME, is now largely the province of amateur radio enthusiasts, who continue to reap the benefits of Operation Moon Bounce and ultimately of Project Diana.

✳ ✳ ✳

The Army's apparent reluctance to publicize Project Diana had long puzzled me. Based more on my own inferences from my father's comments than on anything he actually said, I generally interpreted it as retribution for Jack DeWitt's lack of complete candor about what he was doing, forcing the Army to play catch-up after the fact.

Still, this explanation never entirely satisfied me. Why should those who left the Signal Corps (that is to say, those over whom the Army no longer maintained control) be selectively denied media attention? Why should my father, even decades later, have trouble gaining access to his own earlier work? Even granting that DeWitt was an ask-for-forgiveness-not-for-permission kind of guy working in an organization that prized discipline, it all seemed—well, a bit of an overreaction on the Army's part.

The Eliot article has given me a somewhat different perspective on this issue. In short, it suggests that the Army discouraged media attention less because of what the Project Diana team did wrong and more because of what it did *right*.

The success of PAMOR clearly depended on the Soviets' remaining unaware that their transmissions were being monitored, a consideration that makes the Army's wish to control information that might provide clues about the extent of its capabilities more understandable. Likewise, severely restricting access to information about PAMOR and its debt to Project Diana based strictly on need-to-know provides a more plausible explanation for classifying it above my father's level of security clearance than an arbitrary determination to curtail access to his own work, even though it did in fact have this (presumably unintended) consequence.

* * *

Did my father know about PAMOR? Given that documents such as the Eliot article weren't declassified until 2014, I tend to doubt it. Had he known, he might have been more philosophical about the modest notice given by the Army to Project Diana's milestone anniversaries.

A TALE OF TWO MOONSHOTS: WHAT PROJECT DIANA TAUGHT US ABOUT THE MOON

O n July 16, 1969, the Apollo 11 spacecraft took flight—the realization of President John F. Kennedy's promise in 1961 not only to land a human crew on the moon within a decade, but also to bring them home safely. Four days later, Neil Armstrong gingerly descended the 10-foot ladder of the lunar landing module, the Eagle, and became the first human to set foot on the moon—taking, as he so famously said, "one small step for [a] man, one giant leap for mankind." A TV camera attached to the Eagle broadcast the images back to earth, and the effect was electrifying, dispelling most doubts about the wisdom of devoting so many resources to life support that, by using vehicles without a human crew, might have been spent on the pursuit of pure science.

Almost immediately, Armstrong set about collecting samples of "moon dust" from multiple locations. Twenty minutes later, he was joined by his fellow astronaut Edwin "Buzz" Aldrin, who also began collecting samples. In all, two and a half hours were spent collecting 2.2 pounds of material. They also deployed a seismometer to detect possible tremors and an optical reflector used to measure the distance between Earth and Moon with even greater accuracy. To mark the symbolic importance of the achievement, they planted an American flag along with a plaque that read, "Here men from the planet Earth first set foot on the moon.... We come in peace for all mankind."

The two then climbed back into the Eagle, napped for a few hours, then rejoined pilot Michael Collins in the Columbia command module to begin their journey home. On July 24—eight days, three hours, 18 minutes and 35 seconds after liftoff—the mission ended in splashdown in the Pacific, where the three astronauts were recovered by the USS Hornet. They then remained in quarantine for three weeks, just to be sure they hadn't brought back any extraterrestrial microorganisms, before they were allowed to return to their homes, and to the ticker tape parades that awaited them.

Between 1969 and 1972 there were five more successful moon landings, during which ten more astronauts walked on the moon. One more crewed mission (Apollo 13) orbited the moon in 1970 without landing and returned under harrowing circumstances after an explosion in an oxygen tank crippled the spacecraft, riveting public attention and leading to a radio transmission even more memorable than "one giant leap for mankind": "Houston, we've had a problem" (commonly misquoted as in the 1995 movie *Apollo 13*, "Houston, we have a problem").

No one has been back to the moon since 1972.

* * *

In 2019, *National Geographic* issued a glossy publication titled *The Moon: Our Lunar Companion*, commemorating the 50th anniversary of the Apollo 11 mission. The volume begins with words about the sway the moon has always held over the human imagination and culminates in a detailed look at the first moon-walk on July 20, 1969, the succeeding three and one half years of human exploration, and their enduring legacy. The intervening chapters offer an overview of the history of the scientific study of the moon, starting in 1609 with Thomas Harriot's observations with an early version of the telescope.

As I thumbed through its pages, looking for but not finding a reference to that other moon shot just 23 years earlier, it occurred to me to wonder what, if anything, we had learned about the moon from Project Diana. Most contemporary reports on Project Diana focused on the remarkable innovations in the equipment—the transmitter, antenna, and receiver—that made it possible for the project to succeed in bouncing radio waves off the moon when all previous attempts, including one by Project leader Jack DeWitt himself in 1940, had failed. But did we actually learn something about the moon itself that might have served as the basis for the generation of new knowledge? Something that might have helped pave the way for the lunar landing?

* * *

At least two of the men involved with Project Diana were asked to compare their accomplishment with the lunar landing, and both, in their responses, touched on ways in which Project Diana constituted part of Apollo 11's historical background.

In the oral history interview I conducted with him in 1979, my father remarked on the excitement generated by the success of Project Diana and I asked how it compared with Apollo 11. "Well, it was much less than that in magnitude. But philosophically and sociologically, I think it was important, because as far as I know, it really was the first time that man had *in a measurable way* [italics mine] manifested his influence beyond the reaches of the immediate locale of the earth. Indeed, somebody on the moon could have received radio signals that were transmitted from the earth, but this was really the first proof of it."

According to the *History of Army Communications and Electronics at Fort Monmouth, New Jersey, 1917-2007* (published 2008), Walter McAfee also downplayed the comparison in a 1985 interview, stating that the Project Diana scientists were thinking more about the utility of their work in propagating radio waves than in going to the moon. Nevertheless, he added, the experiment improved our ability to measure the distance between the earth and the moon and contributed to our understanding of the nature of the moon's surface.

* * *

Learning something new about the moon itself, of course, wasn't part of Project Diana's mission. Nor—unlike Apollo 11—were the efforts of well over 300,000 workers and a $25 billion budget thrown at the project. Nonetheless, in the process of determining what they needed to know to succeed, the team contributed to what was then known about the moon in at least two respects.

First, as both my father and McAfee noted, by pioneering a technique of active observation that involved reflecting radio waves off the moon and analyzing the reflected signal—radar astronomy—the Project Diana team developed the most accurate method to date for measuring the distance between the earth and the moon. By extension, and of critical importance for the future of space exploration, Project Diana also showed that the distance and velocity of a spacecraft could be determined very accurately from the earth. Even though the Columbia's onboard navigation system would probably have been adequate for that purpose, the ground-based method was still unsurpassed in accuracy and therefore the main method used to track the Apollo 11 spacecraft.

Second, as McAfee suggested, Project Diana was the first view of the surface of the moon to go beyond what had previously been available only by telescope. Although the equipment then in use could not accomplish anything like the sophisticated mapping possible today, it provided the first information, however crude, on the small-scale topography of the moon.

Project Diana's measurements of surface roughness and mapping of shadowed regions played at least a cameo role in the subsequent debates that preceded the Apollo 11 mission, when knowledge of the nature of the moon's surface was critical not only to the success of the mission but to the lives of the astronauts. The Project Diana team had been seriously concerned about reflective properties of the moon that might need to taken into account when using it as a radar target, as discussed almost obsessively by DeWitt and Stodola in their 1949 paper in the *Proceedings of the I.R.E.* If the moon was perfectly smooth, the reflection would be expected to be comparable to light bouncing off a mirror. From what limited astronomical evidence existed, however, they inferred that the terrain would more likely consist of plains and mountains similar to earth—perhaps somewhat rougher due to the lack of water and air to produce "weathering." They used experimental data from tests of a mountainous area near Ellenville, New York, presumably obtained during the development of methodology to distinguish low-flying kamikaze planes from ground clutter, to adjust their signal transmission and reception parameters.

Later, as radar reflections became somewhat more refined, a serious controversy developed, fueled by a theory proposed by Cornell scientist Thomas Gold that the moon was covered by a fine layer of dust or "lunar regolith" that might be deep enough to swallow—like quicksand—an astronaut or even a spaceship. It wasn't until 1966-68 that a series of Surveyor probes sent to investigate the nature of the moon's surface landed successfully and satisfied NASA scientists that the moon's surface was well-enough compacted to support the weight of an astronaut. So Gold was right about the lunar regolith, it turned out, but not about the quicksand (which for some reason was a disproportionate and almost obsessive fear during that era—someone cal-culated that a third of all movies made in the 1960s included at least a passing reference to quicksand). The assumption under which the Project Diana team operated, it turned out, was closer to the truth.

Even so, the astronauts remained understandably nervous about the ability of the moon's surface to support not only them-selves but also their landing module, which was equipped with legs designed to serve as a built-in launchpad. This was, after all, their only option for getting off the moon. The Eagle had to work right the first time, without benefit of a full dress rehearsal, and there was no rescue plan if it failed. The astronauts knew this. (What they didn't know is that in the event they were stranded, all communication with them would be cut to avoid sharing the last words of the doomed duo with a watching world.)

The collection of moon dust by the astronauts, and even the famous photograph of Buzz Aldrin's bootprint, were part of NASA's ongoing investigation of the nature of the lunar sur-face and its weight-bearing capacity. Fortunately, the astronauts were not swallowed alive by lunar quicksand, the Eagle minus its legs was able to lift off, and as someone wryly put it, "this fear was finally relegated to [the] scientific dustbin of history."

❉ ❉ ❉

In addition to contributing to our store of practical informa-tion about the distance to and the surface of the moon, if only as a byproduct of meeting the needs of their own mission to hit it using radar, Project Diana (as my father recognized) achieved a public relations coup when DeWitt decided to focus on the moon as a target for their efforts, presaging Kennedy's decision to focus on human spaceflight rather than to send robots or dogs to the moon. Both decisions galvanized the imagination and pro-vided a major stimulus to scientific research.

By overfulfilling his original assignment, DeWitt transformed the project from a military exercise to something much broader in scope and interest, something that captured the world's imagination. "Somehow," the *New York Times* declared, "the moon and all the heavenly bodies become more real...more than a guide to navigators and an inspiration to poets...tangible objects to which we can reach out."

Yes, the success of Project Diana resolved any remaining doubts about whether radio waves could penetrate the ionosphere. Yes, it revolutionized global wireless communication. But more than that, the inspirational and aspirational elements of the project emboldened scientists everywhere to consider such new and exciting possibilities as artificial satellites, space probes, and yes, human spaceflight, in the process garnering widespread public support for such efforts in ways that have been well-documented here and elsewhere.

PART II: KING STODOLA, THE RIGHT MAN FOR THE MOMENT

A. THE KINGS AND THE STODOLAS

A VISIT TO LITTLE GLEMHAM

my great grandfather Arthur King

My father's maternal grandfather, Arthur King, emigrated to the U.S. in the late 19th century from Little Glemham, Suffolk, England, where his father and then his brother had served as clergymen at St. Andrew's Church. He left behind a large and close family of siblings, and for the rest of his life, black-and-white photos and letters written double-sided in a spidery script on flimsy "airmail" paper flowed freely across the Atlantic.

At a family wedding a few years ago, surrounded by the people I love best, I had the sudden insight that as the oldest grandchild of Edwin and Beatrice King Stodola, I am the Stodola family matriarch.

With great honor comes great responsibility! In particular, I find myself heir to most of the Stodola family archives. Some of it I've had for as long as I can remember, and some of it came from my siblings, but a great deal of it my father packed away in cardboard cartons when we moved from New Jersey to New York, never to open them again after my mother died. Later the boxes, still unopened, were carted from our basement on Long Island to a storage unit in Florida. My stepmother tried valiantly over time to identify the contents and get things into the hands of the right Stodola child (mostly me, because she knew I would care and share), But like me, she was hampered by cryptic labels (my favorite: "him and me") and lack of dates, and in addition, knew far less about our family history than I.

Although I have eight great grandparents and sixteen great great grandparents, just like everyone else, the Kings have always seemed a little larger than life to me because I heard so much about them from my grandmother, who as an adored only child maintained close ties with her father's relatives in England; and because the Kings were a prolific and retentive lot, leaving a rather large paper burden behind for their descendants to sift through. Over the years, I have threaded my way through most of the documents in my possession and succeeded in identifying many though not all of the photos. I have also connected, through DNA matching and more traditional methods, with second and third cousins who still live in the UK and are much more steeped in King history than I.

* * *

In 1998, before genealogy tourism was even really a thing, Ovide and I made a fascinating visit to the tiny town of La Copechagnière near the Loire Valley of France, birthplace of his ancestor Paul Vachon, a notary public who emigrated to New France in the 17th century. As a family history buff with a keener interest in how our forebears participated in the larger sweep of human history than in an exhaustively-documented series of begats, I concluded that if one hopes to learn about the life and times of one's ancestors, there is no substitute for walking where they once walked.

I vowed on the spot that we would one day make a similar pilgrimage to Little Glemham.

"One day" finally arrived more than 20 years later, in April 2019, when we embarked on a two-week tour of London and environs that included an exploration of Little Glemham in Suffolk and a visit with two second-cousins-once-removed in Sussex. Except for the stress of driving on the left, along narrow roads with many roundabouts, which fell solely upon Ovide, and the stress of navigating, which was my bailiwick, our vacation could only be described as idyllic. Even the weather cooperated—we never even unpacked our umbrellas.

The rest of this essay is about our trip to the UK.

❊ ❊ ❊

In case you are wondering where to find Little Glemham, population 187, it's located just a couple of miles south of Great Glemham, population 224. Even the Brits have to Google to find it. And yes, it's really spelled that way—two m's, no n's.

When we started planning our itinerary, we realized we could include Easter in our schedule, which seemed a propitious time to visit St. Andrew's. I emailed the current rector, who assured me that although St Andrew's is now part of a "benefice" of eight churches—necessary because attendance had dwindled too much to justify weekly services at each church—an Easter service was indeed planned for St. Andrew's.

St. Andrew's, Little Glemham, Suffolk, UK

On our first full day in East Anglia, knowing we might not have an opportunity to poke around much on Easter Day, we stopped by St. Andrew's, which looked exactly as I remembered it in my photographs. The only problem was, it also looked like every other church in every other nearby village, even to the little gatehouse in front, with only minor variations in size and layout. These parish churches date back to the Middle Ages—starting life as Roman Catholic churches and after Henry VIII becoming Anglican—and I guess having hit on a successful formula, the builders decided to stay with it.

We were greeted by two devoted volunteer caretakers, Rod and June Clare, who were busily cleaning and decorating the church for Easter. No need to knock—like all the little churches we visited it is open 24/7, with signage apologizing profusely if for any reason it might have to be closed for even a few hours, just please close the door when you leave to keep the birds out—and we were welcome to stay as long as we wished.

St. Andrew's has been fortunate in its history and is happily well-loved by those few who remain on its rolls. Since it was first built in the 12th century, it has benefited from several updates in its first few centuries, from a restoration project in the 1850s, and from extensive recent repairs. Though the departure of the lead bellringer several years ago led to a silencing of the bells, they can now be heard once again from time to time, thanks to a troupe of ringers that circulates among the local churches.

After we finished our tour of the sanctuary, Rod led us to the King family plot in the churchyard, where we found monuments marking the graves of my great-great-grandfather Richard Henry King II and several of his descendants. Though the stone was partially effaced by time, we could make out the words "parish priest" faintly etched on the side. My grandmother had always referred to him as "rector," so what was that about? Upon consulting Professor Google, I found that rector, vicar, and curate are all forms of parish priest, with the rector sitting at the top of the heap, so no inconsistency after all. (My brother, who has spent lots of time in the UK, claims that "understanding the ins and outs of the Anglican Church is like cricket—if you weren't raised with it, it will always be a Mystery!")

I assumed at first that the charming old home next to the church was where my great grandfather had grown up but later learned that by the time the Kings came on the scene, the rectory had been relocated about a mile away, to a house large enough to accommodate Richard Henry II, his wife Fanny, and their twelve children. A sign in front of that building (now privately owned) identifies it as the "Old Rectory", presumably making the one next to the church the old Old Rectory.

* * *

We arrived early on Easter Day so that we would be sure to find seats. We needn't have worried—the six of us who attended could each have occupied a whole row with room to spare. I was repeatedly struck by the thought that Little Glemham was once home to enough parishioners being born, getting married, suffering, dying, and attending church each and every Sunday to provide Richard Henry King II with year-round full-time employment. He would probably roll over in his grave to learn that the attendees at the 2019 Easter service barely outnumbered the clergy, largely because Ovide and I were there and another man was in town for his father's funeral, and that the current rector is a woman—who rushed off at the end to conduct another service in another part of her benefice.

Also in honor of Easter Day, Glemham Hall (more properly Little Glemham Hall) was open for a tour of the premises. Such tours are common in the UK, as many of the beautiful old Elizabethan stately homes that once formed the backbone of England's social, economic, and governmental system struggle to maintain themselves in a state of "arrested decay." Today the key employees are more likely to be event planners than butlers. Unlike many such tours, however, this one was conducted by the lord of the manor himself, who had grown up in those 80-some rooms and knew it as no docent ever could.

The house was built by the DeGlemham family in the mid 16th century, replacing the moated manor house their forebears had built on the site in the 13th century. In 1709 the North family purchased the property, along with the lordship of the manor, and shortly thereafter made major structural changes to give it the beautiful Georgian facade it boasts today. During the latter half of the 19th century, when my great great grandfather was rector of St Andrew's, the mansion was occupied by Alexander George Dickson, a Conservative Member of Parliament and second husband of the widow of Lord North.

Think *Upstairs Downstairs*. Think *Downton Abbey*. Think (as did I) of the rector of St Andrew's being honored by an occasional invitation to tea at Glemham Hall.

The current lord of the manor is Major Philip Hope-Cobbold, a descendant on his mother's side of the Cobbold family, who made their fortune in the 18th century by founding a major brewery. The Cobbolds bought the house from what was left of the North family in 1923, so in fact Philip Hope-Cobbold's forebears just barely overlapped with the tenure of my own forebears. Still, the tour was both intimate and amusing, and Philip himself was totally charming, leaving us satisfied that we had gained at least a little insight into a social system that somehow allowed the Kings in their rectory to interact in a carefully choreographed way, friendly but at a distance, with the occupants of the nearby manor house.

* * *

As luck would have it, East Anglia was not only the birthplace of my great grandfather, it also played a critical role in the history of radar during World War. The Bawdsey Radar Transmitter Block, just 15 miles from Little Glemham in the village of Bawdsey on the Suffolk coast, was the first operational radar station in the world, where British scientists and engineers secretly gathered during the 1930s to demonstrate that radio waves could in fact be used to locate moving targets. Chain Home, code name for a series of early warning radar antennas strategically placed all along the British coastline to detect and track incoming aircraft, fanned out from Bawdsey.

During the Battle of Britain in 1940, their resources stretched almost to the breaking point, the British sent a delegation to the United States to propose a marriage of British science and knowhow with American industrial capability. Public sentiment in America favored neutrality; Henry Tizard, head of the mission, took the bold step of showing the Americans the technical innovations they had achieved without any promise of reciprocation—including the high-frequency magnetron that, by reducing the size of the antenna required, enabled installation of radar in aircraft. As Tizard hoped, the sheer impact of British superiority in the development of radar was sufficient to convince the Americans it was in their own best interest to support the British effort, and thus began the amazingly productive British-American collaboration in the development of radar.

It was just around this time that my father left his entry-level job at the War Department assigning radio frequencies to Army facilities ("boring!") to begin his fledgling career as a radar scientist. Although the Tizard Commission visited Bell Labs in New Jersey and Columbia University in New York City, I have found no evidence of their having stopped in Belmar—and in fact, the subsequent locus of collaboration focused on the creation and development of the famous Radiation Laboratory at the Massachusetts Institute of Technology rather than on work already in progress by the Army Signal Corps at Camp Evans. Still, it seems likely the American commitment to the British war effort, cemented by the Tizard Commission, set the stage for my father's career in radar research and his particular expertise in moving target detection.

Although we didn't have a chance to visit the Bawdsey Radar Museum, we managed to spend a couple of engrossing hours at the Parham Airfield Museum near Little Glemham, housed in the original World War II Control Tower of Framlingham Air Force Station #153. By way of background, if the American military was to lend its might to the war effort, it needed bases of operations within range of German cities and factories. In the UK alone, over 200 airfields were built for or transferred to American troops during that era. Each was designed to house 2,500 troops, which often dwarfed the population of the little villages that hosted them. All told, over 2,000,000 American servicemen were stationed in Britain at some point during World War II.

The Parham Airfield was one such facility, built in 1943 and immediately turned over to the 390th Bombardment Group. This unit carried out more than 300 combat missions in the Boeing B17 "Flying Fortress," during which 19,000 tons of bombs were dropped and 342 enemy aircraft were downed. Nearly 200 American planes never returned, and more than 700 American service members were killed in these risky missions. Also worthy of mention are the humanitarian flights undertaken by the Americans just before V-E Day to supply desperately-needed food to the Dutch.

In a world where Americans aren't universally welcomed or appreciated, it was heart-warming to bask in the affection and gratitude with which the Yanks are still, even after all these dec-ades, remembered at Parham.

* * *

During the year or so before our trip, I had become friendly with my DNA match Jane, a second-cousin-once-removed who like me is descended from Richard Henry II, so we spent a few days at an inn near her home in Sussex, south of London. Jane and her husband David turned out to be the most gracious of hosts imaginable. Jane is an excellent cook who served up steak and kidney pie and other traditional delicacies, and David en-tertained us with a video he had made for the BBC back in the 1980s in which he persuaded some friends to help him mow his lawn by staging a lawnmower tango worthy of Monty Python. We also watched an episode of *Escape to the Country*, the BBC ver-sion of *House Hunters* (but much better), featuring a visit to the one-of-a-kind Black Cow Pure Milk Vodka distillery developed by a well-known maker of cheddar cheese in West Dorset—where their daughter (my third cousin) happened to be employed.

Cousin Jane introduced me to her cousin (and like Jane, my second-cousin-once-removed), Ian, and we enjoyed a delightful luncheon with him and his wife Nathalie. Ian, unlike me, is a bona fide genealogist, so it was gratifying to be able to help him fill in the blanks on Arthur's family (including four generations of descendants with the middle name of King).

Of all the things he shared with me, nothing was more fascinating to me than his photos of the King Boys' Butterfly Collection. One of the few pieces of information I could coax from my father about his relationship with his grandparents was his fond memory of butterfly-collecting expeditions with his grandfather Arthur. Thanks to Cousin Ian, I now under-stand that this activity was not just an idiosyncratic passion of Arthur's, it was part of a King family tradition that he must have hoped my father would enjoy and pass along to the next generation. (In fact, it was not just a King family tradition, it was, as someone said, "a stock hobby of English vicars" and enjoyed by many prominent Victorians including Chamberlain and Churchill. It has more than fallen out of fashion, it is now pretty much socially unacceptable—because it obviously doesn't end well for the butterflies. My father never took his children butterfly hunting!)

the King Boys' Butterfly Collect (one of three trays)

After our time with Jane and Ian, we had one day left to tour Sussex and decided to spend it exploring Canterbury and its Cathedral. Coincidentally, the date of our visit was probably very close to those April days more than six centuries before on which Chaucer's fictional pilgrims were busy concocting tales to entertain their fellow adventurers as they wended their way towards the Canterbury Cathedral. Even from today's perspective, the cathedrals of the Middle Ages are amazing structures, but to appreciate them fully, it is necessary to imagine them rising up almost literally out of nowhere, with nothing nearby of anywhere near the same magnitude; and then to imagine yourself a penitent who has never been more than a few miles from where you were born, whose sole experience with churches is with something on the scale of St. Andrew's in Little Glemham.

No wonder the pilgrims felt themselves in the presence of something supernatural and otherworldly.

❋ ❋ ❋

My grandmother's stories had somehow given me the impression that a long line of Kings had been rectors of St. Andrew's in Little Glemham since time immemorial. In actual fact, my second great grandfather Richard Henry King II appears to have been the first to serve in that capacity, succeeded by his son Edward Septimus King (younger brother of my great grandfather Arthur, and of Richard Henry III, grandfather of my cousins Jane and Ian).

I have no further evidence of a King dynasty in Little Glemham. Richard Henry II himself was in fact baptized not in Little Glemham but in Mortlake, south of London, where the King family had apparently resided for a very long time: The King family crest that my cousins regard as authentic (as opposed to a somewhat different version painted by my grandmother that was a prominent part of my childhood iconography) is labeled "King of Mortlake/Arms granted 1589." I don't know the profession of Richard Henry King I, my third great grandfather, but like his forebears, he was buried in Mortlake. His father, my fourth great grandfather Dr. Charles King II, was a physician who lived and died in Mortlake.

For much of its history, Mortlake ("mort" apparently meaning "salmon," not "dead") was officially a village, though a village large enough to support a variety of industries including potteries whose products are still very much in demand, a tapestry works, a sugar refinery, and breweries at various times in its history. Its dreamy beauty was captured by J. M. W. Turner in two landscapes painted in 1826 and 1827, depicting views of and from a large townhouse then known as Mortlake Terrace, commissioned by its owner. Currently, Mortlake is a suburban district of London and a popular sleeper community.

In Little Glemham it was possible to walk where my ancestors had walked because the area has retained its small-village character and hasn't changed beyond recognition. Had we tried to add Mortlake to our itinerary, we would have faced a more formidable challenge.

At any rate, a quest for another day.

ON BEING NAMED "KING"

T he untimely death of Prince in 2016 caused me to reflect on my father.

This was not because of even the slightest overlap in aesthetic preferences. My father's musical tastes ran the gamut from Gilbert and Sullivan to John Philip Sousa. He was the kind of guy who enjoyed going to a high school band concert even if his own kids weren't in it. Not your typical Prince fan.

No, the strictly tangential link between Prince and my Dad is that both had unusual given names, names that implied a royal connection that each in his own way ended up earning.

In my father's case, "King" was not his first name—*that* was "Edwin," after his father; but "King" was his mother's birth name and no one ever called him anything but "King"—nor did my father ever seem the least bit abashed by it (unlike Prince, whose name repeatedly got in his way). Consequently, it never struck me as being at all out of the ordinary, though some of my friends later told me how odd it seemed to hear a man addressed as "King" as casually as their own fathers were called "Jim" or "Ken."

Later, as I sifted through the extensive collection of King family letters and documents I inherited from my father, I came to appreciate my grandmother's love for her family and pride in their name. Her father, Arthur King, had left Little Glemham and sailed to Philadelphia in 1885, presumably seeking economic opportunity. (What Arthur actually did for a living I do not know, though at the age of 20 he was employed, according to the census, as a "Clerk in Merchant's Office" in Bowdon, Cheshire.) In January of 1889 he married a Pennsylvania girl named Vergetta Jane Sayers, and ten months later their adored only child Beatrice was born. Arthur was naturalized as a U.S. citizen in 1893—an act requiring him to forswear allegiance to Queen Victoria.

Despite his love for his adopted homeland, however, Arthur maintained strong ties with his family in Little Glemham for the rest of his life, returning frequently to visit them. My grandmother corresponded regularly with her first cousins in England. So my grandmother's King heritage was, for her, living history. She always called herself Beatrice King Stodola and sometimes, professionally, Beatrice King.

❊ ❊ ❊

Luckily for my grandparents and for Prince's parents, the U.S. is quite tolerant of obscure or unusual name choices. Had they been born elsewhere, these names might have been rejected out of hand. "Prince" and "King," for example, are both near the top of a no-list maintained by the New Zealand government for violating the rule that "acceptable names...should not resemble an official title and rank." (Ironically, the name "King does not necessarily imply descent from royalty but rather refers to a man with a kingly bearing, or to someone who played the part in a pageant or earned the title in a tournament.)

Since my father, no one else in the family has ever been known as "King." The tradition of using King as a middle name is now in its fifth generation, however, so I guess the Kings will continue to be with us for quite a while!

WHEN EDWIN
MET BEATRICE

cover of flyer offering instruction in elocution (Beatrice)
and piano (Edwin)

T he history of the Stodola family in the United States began with my great great grandparents, who were part of a wave of German immigration in the mid 19th century in response to war and turmoil in Europe. A subset, like Henry and Barbara Stodola, were Ashkenazi Jews who in addition to fleeing civil unrest were seeking to escape antisemitic persecution. Hy (as he was called—likely derived from Chaim) arrived in Boston from Prussia on July 21, 1849, and then made his way to New York City. Barbara arrived a few years later; they must have met and married in New York. Their first child, Malia, was born in 1955, their second, my great grandfather Joseph, was born in 1858, followed by Lena in 1860 and Samuel in 1862. According to the 1870 census, Hy was a baker. Barbara died a widow in 1908.

On May 29, 1881, my great grandfather Joseph, a dry goods merchant, married Bertha Eilau. Joseph and Bertha had three children, Gilbert Isaiah (born in 1883), my grandfather Edwin Sidney (born in 1885), and Ruth, the baby (born in 1892). Gilbert listed himself as an editorial assistant on his World War I draft card but also stood as proof of a scientific bent in the family, having published an article in the January, 1923 edition of the *Scientific American* about an ingenious device he'd invented for installation in taxicabs to allow the fleet owner to determine the time and distance covered in each trip. So my father's engineering abilities and love of gadgets, though they probably puzzled his artistic parents, apparently didn't come from out of nowhere.

My grandfather must have displayed a gift for music early in life and was evidently encouraged to pursue it. By the time he was in his late teens or early twenties, he had achieved critical acclaim as a concert artist, including the obligatory Carnegie Hall recital, and was also working as a piano teacher. His advanced training included six years under the tutelage of Henry Holden Huss—largely forgotten now, but in his day a well-known American pianist and composer; performances of his pieces can still be readily heard on YouTube. They evince a conservative taste in music and are quite free of avant grade rhythms and dissonances. (It is worth remembering that Brahms lived until 1897, when my grandfather was age 12 and Henry was 35, so the romantic tradition was hardly ancient history during that era.)

I'm not sure when these six years of study with Huss began and ended. My family archives includes a tattered photo of Henry Holden Huss and his wife, the American soprano Hildegard Hoffmann (clearly a marriage made in HHHeaven!). On the other side, the photo is inscribed "To my dear pupil Edwin Stodola" and dated June '04—likely a token or souvenir of Henry and Hildegard's marriage, which took place in that month and year. So—the six years my grandfather studied with him must have included if not started in 1904.

At the end of his training with Huss—shall we guess around 1910?—my grandfather left his native New York to join the faculty of a conservatory in St. Joseph, Missouri. It was sometime during this period that he met my grandmother, a native of Missouri. Did they become acquainted after he arrived there, or did he meet her elsewhere and move there to be near her? I cannot say.

Florence Beatrice King, the doted-upon only child of Arthur and Vergetta Sayers King, was born in 1889 in Savannah, Missouri. Like Edwin, she found her calling early in life and participated in theatrical performances as a child. Her training was in a profession that almost doesn't exist today, as an elocutionist or *diseuse*; perhaps the closest modern equivalent would be a performance artist. After finishing her secondary education she attended the Chicago Conservatory of Music and Dramatic Art, followed by postgraduate work at the Columbia College of Expression and the American School of Expression and Oratory. She then returned to Missouri and coached students in drama and public speaking. She also painted and wrote poetry and plays. She was sociable and effervescent and had a wide circle of friends. She must have dazzled the quiet Edwin.

Edwin and Beatrice married in St. Joseph on January 11, 1913. It was clearly a love match, but one that came at great personal cost. When he proposed to Beatrice, her parents, after duly inquiring about his moral character—probably of Henry Holden Huss!—welcomed him into their family. His own parents, however, were unable to accept his marriage to a gentile, and sadly all contact with them ceased.

One of my most treasured souvenirs is a charming brochure assembled by Beatrice and Edwin, apparently to help in recruiting students. This brochure, with a "Beatrice" column and an "Edwin" column, is actually the source of almost everything I know about their professional training. Since the brochure lists my grandmother's name as Beatrice King Stodola, it must have been published shortly after their wedding. (In fact, it might have replaced an earlier version that did not include my grandmother's married name.)

Not long after, their Missouri saga came to an end, and by the time my father was born in October of 1914, they were living in Brooklyn. Now that my grandfather was back in New York, I wish I could report a reconciliation with his parents, but alas, that did not happen, then or ever. Edwin's brother Gilbert, so far as I can tell, never married, and his sister Ruth, marrying late in life, never had children. So my father and (later) his two younger brothers were Joseph and Bertha's only grandchildren. My grandmother made a couple of valiant attempts to visit her in-laws, hoping the sight of their adorable little grandson would melt their hearts, but she was not invited in. Only now, through the miracle of ancestry.com, am I slowly piecing together my grandfather's story and even tracking down a couple of third cousins on the Stodola side.

Upon returning to New York City along with his bride, my grandfather resumed his association with Henry Holden Huss and Hildegard Hoffman Huss in the capacity of "pupil-assistant" to both, which included serving as an accompanist for their other students.

This ushered in what might be called the Golden Age of my grandfather's career, as a Google search turns up several reviews of concerts and recitals in which he played pieces I could only dream of tackling. The oboist and musicologist Lisa Kozenko, in a doctoral dissertation focused on this era, describes "a musical reception in honor of the American baritone David Bispham (1857-1921) along with the Husses. The guests included the

Vincent Astors, Carolyn Beebe, Elizabeth Sprague Coolidge, the Hermann Irions, Hugo Kortschak, Mrs. Ethelbert Nevin, and others. In its April 7, 1917 issue, *Musical America* reported that Huss was asked to improvise ('A thing which he does fascinatingly'). His student Edwin Stodola suggested "D B" as [a theme honoring] Bispham: 'It was on that that Mr. Huss built his splendid improvisation, which won warm favor from all present.'"

So we have now arrived at the May 6, 1918 concert in Rumford Hall and my grandparents' subsequent move to Boston, as described in my essay on the 1918 pandemic elsewhere in this volume. Although I concluded in that essay that they had not moved to escape the pandemic, they almost certainly also did not move to advance Edwin's career as a concert artist; if anything, probably the opposite. Although Boston was once the hub of the national chamber music scene, the center of gravity had by now shifted to New York City. Had he been bent on pursuing a concertizing career at the exclusion of all else, my grandfather should have stayed put and remained under the wing of the Husses.

"Being a musician is not an easy profession in any era," wrote Kozenko in her dissertation. "[By 1920], the competition from radio, recordings, jukeboxes and movies confronted anyone attempting to pursue a career in music." The May 6 concert appears to have been almost a swan song. With a growing family and the 1918 influenza pandemic emptying concert halls, he had likely decided it was time to shelve the dream of supporting his family as a musician and move on to a different life plan.

This is not to say that he abandoned music altogether. Performances in Boston included a musical accompaniment to a lecture on "The Relationship of Poetry and Music" by my grandmother at the Boston Public Library in April 1922; piano music in a radio broadcast in October 1922, "Rhymes and Music for Little Folks," also featuring readings by my grandmother. Similar notices appeared occasionally in the *Kingsport Times* during their few years in Tennessee, where he ran a print shop. His 1957 obituary states that in addition to teaching and concertizing, he served as director of music for the [New York] Veterans Administration, where he assisted in the training of veterans studying music at Columbia University, Juilliard, City College, and New York University.

But there is nothing to suggest that he ever gave up his day jobs, in which he used his keyboard skills as a "typewriter" and a printer. Music had become an avocation.

* * *

Despite the disappointments that beset them, theirs was a happy union. My grandfather was a gentle man and a gentleman, quiet and reserved, but he never passed my grandmother's chair without touching her shoulder. He even enjoyed "helping" her with her garden, which mostly consisted of planting seeds and pulling weeds under her direction.

By the time I knew him, his hands were rusty with arthritis and disuse, and he had grown hard of hearing. He never willingly played the piano in my presence, and when he was cajoled by my grandmother into accompanying family musicales, it was clear to me even as a child that his skills had sadly atrophied. Yet so far as I know, he was content with his lot.

B. THE BOY, THE MAN, THE DAD

KING STODOLA:
A BRIEF BIO

my dad with his grandparents, Arthur and Vergie King

My father was born in Brooklyn, New York on October 31, 1914. (For that reason, Hallowe'en was always a happy holiday in our house. Costumes, trick-or-treating, all that candy—plus cake, ice cream, and rousing choruses of "Happy Birthday to You!" What could be better?) His parents, Edwin and Beatrice King Stodola, were professional performing artists who often gave joint concerts featuring classical piano pieces by Edwin and dramatic readings by Beatrice. How surprised they must have been by their eldest son's strong bent for math and science! But apparently they took it in their stride and encouraged him to follow his lights where they led him.

In 1918, my grandfather accepted a position in Boston, and between the ages of four and eleven, my dad lived with his family in a series of rental homes and apartments in a cluster of Boston suburbs—Randolph, Milton, Wellesley, Brookline, Auburndale. Shortly after they arrived, his brother Quentin was born. Two years later, in 1920, the family welcomed the last of the three Stodola boys, Sidney. (In a burst of enthusiasm for the newly elected incoming president, my grandparents gave Sid the middle name of Harding, a decision they later came to rue.)

My dad's maternal grandparents, Arthur and Vergie King, joined them in the Boston area sometime between 1920, when the census showed them as living in Lakewood, New Jersey, and 1924, when they were ensconced in a flat at 5 Waban Street in Wellesley.

Presumably it was during this period of their lives that my grandparents discovered and fell in love with Cape Cod. They later bought and refurbished a small cottage in Wellfleet that my grandmother christened Shining Sands, where my sisters and I spent many idyllic summer vacations during our childhood. It was loaded with rustic charm back then; my grandmother would be amazed to see it valued on Zillow at nearly $1 million—much "improved", of course.

It was hard to get my father to talk about his childhood. During my oral history interview with him in 1979, I had to probe repeatedly to elicit much in the way of detail. Questions about where he came from were usually met with his stock response, "I'm a native of wherever I'm living now." For a man who claimed to remember little of his childhood, however, he managed to come up with quite a few touching or amusing anecdotes about his days in Massachusetts.

His earliest memories were of a dog named Sandy, a St. Bernard-Newfoundland mix they adopted in Milton. Later the family moved to the larger and more urban town of Wellesley, where after attending the Hunnewell Elementary school, my father was transferred to a rapid-promotion class in a Wellesley Hills school that enabled him to complete three years' work in two. Sandy's Wellesley career was less distinguished—he was jailed for fighting with the town clerk's dog.

King, Quent, and Sid with their dad and the incorrigible Sandy

Knowing of their enthusiasm for butterflies, a neighbor named Mr. Denton sold my dad and his grandfather cocoons for a nickel apiece. Mr. Denton was also an avid fisherman and took my father fishing in his canoe. Another neighbor, Thomas Reed, was a craftsman who fashioned a washing machine from a butter tub and encouraged my father's interest in learning how things worked—which he pursued by walking several miles to the Waltham Woolworth to buy electrical gadgets.

My father's well-to-do friends the Wentworths invited him to their summer home near Lake Massapoag in Sharon, where he and the Wentworth children made ice cream in a hand-operated freezer and enjoyed songfests at a nearby Salvation Army camp. The Wentworths's maid once scolded the children for too much rough-housing—only to discover there had been an earthquake!

In 1925, his grandfather Arthur died. Shortly afterward, his father landed a position as training director for the *Kingsport Press* in Kingsport, Tennessee, an offer he apparently couldn't refuse. The whole family, including Sandy the dog, piled into their Model T touring car and laboriously plowed their way south along potholed roads coated with Kentucky red mud.

And so the family's Boston idyl came to an end—and along with it, my father's carefree childhood. His immediate future held a move to junior high school and the turbulent days of pre-adolescence, compounded by a fair dose of culture shock. For him and his younger brothers, there would be no more rubbing elbows with the literati.

A year later Vergie died in Kingsport, either during a visit or more likely because—now a widow with no other ties to Massachusetts—she accompanied her only daughter when the family moved. Vergie's remains were returned to Wellesley for burial alongside her husband.

Beatrice with King, Sid, Quent, & Sandy in Kingsport

Kingsport in the 1920s was something of a boomtown, boasting the Eastman Chemical Company and a large paper mill in addition to the *Kingsport Press*. After gaining a foothold there, my grandfather was able to set up his own multigraph business, where my father helped him with printing, typography, and composition. My father also learned to drive during the Kingsport years, when he was so small that he had to sit on a pile of telephone books to see through the windshield.

But as my father told it, Kingsport also had a darker side: The Ku Klux Klan was active, and the men carried revolvers in their coat pockets. My grandmother attended a small Episcopal church in a YMCA that was regarded with suspicion because it was considered to be just short of Roman Catholicism.

The Stodolas missed the liberal spirit they found congenial and the cultural amenities they had enjoyed in New York and then in Massachusetts. They remained in Kingsport through my father's junior high school years, then returned to New York in time for him to attend Brooklyn Technical High School, and for the younger boys to take advantage of New York City's superior educational system.

After graduating from Brooklyn Tech, my father went on to attend the Cooper Union, an engineering and art school noted at the time for being highly selective and (importantly) tuition-free. He made many lifelong friendships during this era, both at school and among the Unitarians whose meetings he attended (in preference to the Episcopalian church to which his mother belonged). Even then he understood that friendships need nourishment if they are to last. I think particularly of his high school classmate Karl Dworak—of the efforts they made to get together with their wives and children and the fun they had carrying out amusing little science projects when they did. (The Dworaks had four sons, including an acrobatic pair of twins who were kept in two playpens, one upended on top of the other. The twins didn't seemed to mind!)

On July 22, 1939, he and my mother, Elsa Dahart, were married in the All Souls Unitarian Church in Washington, DC. At the time he was working for the War Department assigning radio frequencies for Army facilities. Later, seeking a more research-oriented job, he accepted a position in signal detection and the development of radar at Fort Hancock in Sandy Hook, New Jersey, and my parents relocated to the Jersey Shore town of Long Branch. After Pearl Harbor, when the country moved to a wartime footing, his work was transferred to the Evans Laboratory in Belmar and my parents moved to nearby Neptune.

I was born in Neptune in February of 1943, the eldest of their four children. I like to think my parents named me "Cynthia," the Greek version of Diana, goddess of the moon, in a moment of prescience, though in truth it seems unlikely that Project Diana was on anyone's radar as soon as 1943. Although we were not a demonstrative family, I always felt that I was valued and cherished. The nurturing environment, however, was probably more my mother's doing. My father tended to relate to children as miniature adults.

Shark River Hills, an "unincorporated community" within Neptune Township, started out as a beach resort, with a lot of homes standing empty all winter. Once World War II was underway, however, the year-round population burgeoned with both military and civilian personnel assigned to Camp Evans, and my family was part of that transition. Today it is all paved roads and marinas and lawns, but when I was little it was very rural—not agricultural, but mostly gravel roads where you could ride your bike and not pass any houses or see any people for long stretches. Yards were mostly sand and rock gardens; you couldn't have a lawn unless you hauled in loads of topsoil, which only a few of our neighbors did at that time.

When they first moved from Long Branch to Shark River Hills, my parents rented a tiny house on Clinton Place, down near the Shark River, from a couple who brought in a little extra income by vacating their home and residing in an apartment over their garage. This is where we lived at the time of Project Diana.

I wish I could give a firsthand account of that event, which my father always regarded as his proudest achievement; but although I have many memories from that era of my life, including the birth of my sister Leslie in 1945 and fleeting glimpses of the house we lived in, the excitement surrounding the successful moon shot isn't one of them. The late David Mofenson, my earliest playmate and fellow Project Diana legacy (the Signal Lab families tended to socialize mainly with each other during those early years in Neptune), recalled these moments vividly, so perhaps this isn't the sort of thing little girls tend to hang onto; in any case, not this little girl. The only recollections I have of my dad's work are those tedious interludes in the car with my mom and baby Leslie outside those old Marconi Lab buildings, waiting for him to emerge. We never went in.

In 1947, my father took job with Reeves Instrument Corporation in New York City and began several years of commuting by rail (including a running bridge game on the way in and reading murder mysteries on the way home), punctuated by frequent business trips around the country. At that point he became a weekend father, leaving my mother to shoulder most of the burdens and pleasures of child-rearing. Still, he was very much present most weekends, often spending his Saturday mornings flat on his back, working on his car. We never needed a plumber or electrician because he was an amazing Mr. Fixit who could take on just about any washing machine or toaster that dared to break. He also became an avid outboard motorboater during these years, an interest that persisted even after we left Neptune. We had a series of vessels with names consisting of the first two letters of the names of his children—first CYLESH (CYnthia, LEslie, SHerry) and later, after RObert was born, CYLESHRO.

At around the same time my father left the Signal Corps to go into private industry, we moved to a comfortable bungalow on Pinewood Drive that, though not large, was baronial compared with the little rental on Clinton Place. It had screened-in porches front and back, where the whole family slept on very hot nights, and where my friends and I played canasta, Old Maid, jacks, pickup sticks, checkers, Parcheesi, Monopoly—and let's not forget the ouija board!— all summer long; and a floor plan that allowed our electric trains to disappear into another room and circle around a wall before reappearing. Not sure who loved those trains more, the children or their father!

In 1956, Reeves Instrument Corporation moved from Manhattan to Garden City, Long Island, and the commute from Neptune was no longer feasible. My dad considered changing jobs at that time, which also would have involved a move (to either California or Michigan), but in the end Reeves made the best offer. And so our Jersey Shore days drew to a close.

After a little house-hunting, my parents realized they could get a better house by living further out on the island—meaning my father would once again be relegated to a 45-minute commute, this time by car. My mother thought most of Long Island was too flat and zeroed in on the town of Northport because it had some hills. Our real estate agent spent a whole day ferrying us around and scoping us out before revealing that there was a house for sale on Stanton Street, right next door to her own home.

It was truly a wonderful house, with a bay window, a Dutch front door, knotty pine paneling, a beautiful tilework fireplace, bedrooms with special features (mine had an outdoor balcony and Leslie's had a loft), and a huge green yard, a novelty for us. We all fell in love with it immediately. A bonus, especially for my dad, was an active Unitarian fellowship located just a few miles away in Huntington; it was the first time we'd ever lived closer than an hour's drive to the nearest Unitarian church. Another bonus was the opportunity to pilot the latest CYLESHRO across the 8-mile expanse of the Long Island Sound (usually including one of his typical yacht-in, yacht-out visits to friends in Connecticut—which turned into fly-in, fly-out visits after my mother died and he took up flying).

My parents' relationship was amazingly harmonious, something enough others have commented on to convince me I'm not selectively forgetting unpleasant moments. I never remember hearing them argue, let alone raise their voices to one another. A running joke between my parents was my father's nickname for my mother, Blondie—not because she was blonde (she wasn't), but because they in some way identified with the comic strip characters Dagwood and Blondie Bumstead—Blondie handing

Dagwood his briefcase and hurrying him into a catapult positioned to propel him out the front door just in time to catch his bus; Dagwood constructing and wolfing down monster three-layered sandwiches with revolting combinations of ingredients; Dagwood providing endless exasperation-cum-amusement for the long-suffering ditzy-but-wise Blondie. My mother, though long-suffering and wise, was never ditzy—but that was part of the joke.

Having watched my father's equally courteous behavior towards his second wife, Rose, whom he married three years after my mother's death in 1965, I think it speaks to his fundamentally gentlemanly nature, though probably also to a proclivity for choosing forbearing women as his life partners. But although both his wives were mild-mannered and softspoken, my father admired high-powered professional women immensely and counted many within his wide circle of friends. It was from him, much more than from my mother, that I received the message that women could do anything and be anything they chose. He strongly urged me to take all the math and science I could cram into my schedule and made sure I enrolled in AP physics and AP calculus, something I've never regretted as it has opened doors that would otherwise have been closed to me.

My father's gadgeteering proclivities led to inventions that resulted in about twenty patents—one of the most important being a long-range tracking system for radars that was widely used in satellite tracking, including Project Mercury. The most lucrative, however, was an electronic thermostat, the Automated Temperature Relation Control System, an updated version of which can still be found in most houses. The patent was infringed upon by Honeywell, which reverse-engineered the product and marketed it on a scale beyond anything Reeves Instrument Corporation could have dreamed of. My father won a court case against Honeywell, resulting in a royalty stream that lasted for many years. He always said he had the best of both worlds with this arrangement and claimed it put me through college.

Sometime during the 1970s, my father's career took a disastrous turn. Back in 1955, Reeves had merged with Dynamics Corporation of America. When the electronics industry fell into a deep slump, the merged operation pulled up stakes and relocated to Florida in hopes of riding out the downturn by lowering expenses. It seemed to work, but not for long, and when DCA went bankrupt, it took Reeves down with it. First to be paid off were the stockholders and creditors, leaving precious little for the senior executives at Reeves, including my father, who lost their pensions.

To eke out a living, he took a series of consulting jobs requiring him to live in seedy motels and tiny furnished flats all over the country, plus a more extended stint near Syracuse, New York—until he finally secured a job in Electronic Warfare at the Pentagon. All his years at Evans Lab had qualified him for a government pension, but the amount was based on his highest three years of salary, the famous federal high-3, and for him those years were 1946, 1947, and 1948. The Pentagon job enabled him to accumulate three more years, enough to provide an adequate retirement income. So his heroic efforts paid off, and in the end he landed on his feet. It wasn't until I myself reached retirement age that I fully understood the gravity of his plight. But although he was perfectly capable of a good rant, I never heard him complain about the injustice of his situation at a time of life when he should be winding down rather than revving up.

The last few years of my father's life were blighted by Alzheimer's Disease. Although it's said that Alzheimer's is harder on the caregiver than the victim, my father was well aware of his deficits for much of that time, and angry and frustrated by his declining abilities and failing memory. Fortunately, his downward spiral was relatively rapid, about four years. By then he had enjoyed a long and mostly fulfilling career; he had lived to see his children reach adulthood and meet his grandchildren. My stepmother, his life companion of more than two decades, was still by his side. Although I devoutly wish we could have enjoyed his company into advanced old age, the arc of his life was by and large complete, and his death in April of 1992, at the age of 77, was actually a mercy and a release (in stunning contrast to my mother, felled in midflight at the peak of her powers). Even with advance directives, no one can speak for others when the moment actually arrives, but I doubt he would have wanted his sad deterioration to be further prolonged. Or so I hope.

PANDEMIC

The annual flu season of 1918 started in the Spring. As flu seasons go, it was relatively mild; those who became ill typically suffered through a several days of chills and fever, then recovered. The number of reported deaths was low. A year earlier, on April 6, 1917, the United States had entered the Great War, and news of major offensives in Europe, not the flu, dominated the headlines. As summer approached, the flu subsided as usual.

Then, on August 11, the Norwegian vessel Bergensfjord docked at a Brooklyn NY pier. Twenty-one of those aboard—10 passengers and 11 crew members—were ill. A team of doctors and city officials, having been warned in advance, were on hand to meet the ship and quickly sent those who were ill to nearby hospitals. The first few cases of what turned out to be one of the deadliest pandemics in human history had landed in New York City.

Despite efforts at containment—the pier was immediately placed under strict quarantine—the disease started to spread, insidiously at first, but soon sweeping through the city's densely-populated neighborhoods like wildfire. Unlike its earlier incarnation, this second wave turned out to be a merciless beast. The first death in New York City was recorded about a month after the Bergensfjord arrived, by the first of October the daily death toll had reached around 50, and by the middle of the month over 400 people were dying each day. October of 1918 was the deadliest month of the entire pandemic. Eventually, over 20,000 New Yorkers lost their lives.

And a singularly unpleasant death it was. The historian Mike Wallace described "patients gasping for breath as their lungs filled with bloody frothy fluid." Within days or even hours, they basically drowned in their own bloody fluids. Victims could awake feeling fine and be dead by midnight.

Consistent with the usual demographic pattern, children under five and adults over 70 were at elevated risk. Unlike most influenza epidemics, however, healthy young adults between the ages of 20 and 40 were also highly susceptible to infection and suffered by far the highest mortality rates, with pregnant women being at the highest risk of dying if infected. Amazingly, except for individuals over 70, older adults, the group upon which the flu usually preys most heavily, were largely spared, producing an unusual W-shaped death curve—likely because some similar but less lethal bug had struck during their childhood and conferred lingering immunity.

* * *

Sometime between May 6, when my grandfather gave a concert in Rumford Hall on Worth Street in lower Manhattan, and September 12, when my grandfather was issued a draft card in Massachusetts (in response to an August, 1918 amendment to the Selective Service Act raising the upper age limit for eligibility), my father and his parents moved from their densely populated Brooklyn neighborhood to the relatively quiet town of Randolph, a suburb of Boston.

During the dark moments of the Covid-19 pandemic in 2020, it occurred to me to wonder whether the Great Influenza played any part in their decision to leave Brooklyn. Historically, even when contagion was less well understood, people realized that crowded conditions might contribute to the spread of disease, and city-dwellers who could afford to do so often fled to less crowded areas when a pandemic struck.

Public health officials anticipated from the start that New York would be severely affected by the Great Influenza. It was a large port city whose population had mushroomed due to recent waves of immigration. Many of its doctors were overseas caring for the wounded in World War I, which at the time appeared to have no end in sight—though in fact it ended rather abruptly with the Armistice on November 11 of 1918. The installation of subways and elevated trains had created a brand new phenomenon, "rush hour." New York was a center for international travel and shipping and the departure point for more than a million troops headed for battlefields in France. In short, New York was a veritable breeding ground for the flu, and Brooklyn was at its center.

If ever a family had targets painted on their backs, it was the Stodolas. Edwin was about to turn 33 and Beatrice was 28 (the modal or peak age for mortality in the 1918 pandemic), her pregnancy putting her and her unborn child in even graver danger. My father, aged three, was also in a high-risk category.

The likelihood of my grandparents having known what was coming in time to stay ahead of the fast-moving tsunami, however, is low. Front-page news focused on the Great War, while reports on the Great Influenza were buried or absent. And not just by chance; countries engaged in fighting actively suppressed news of the outbreak to avoid undermining morale. (Because Spain was not at war, the Spanish press wrote freely about it—hence the misnomer "Spanish flu.") When the approaching infection could no longer be ignored, local officials, abetted by the press, typically reassured the public either that it wasn't really the Spanish flu or that the worst was over. Not until people saw their friends, neighbors, and family members dying before their eyes, dying horribly, did they recognize the disconnect between what they were being told and reality.

If against all odds Edwin and Beatrice were somehow cognizant enough of their plight to entertain the hope that moving to a less densely populated suburban area would improve their chances of surviving the pandemic, they were quite simply mistaken. The Boston area *per se* was hardly a haven from the flu. In fact, it struck the port of Boston even before it arrived in New York; indeed, in the end, the death rate from influenza in Boston was even higher than in New York. Nor did the suburbs provide any refuge. Except for a few communities that were either smart enough or fortunate enough to achieve true isolation from the outside world, there was nowhere to hide from the Great Influenza. In fact, there is no evidence for either Boston or New York that the suburbs fared better than more densely populated urban areas.

Rather, my grandparents' move was probably motivated by a desire for a more suburban lifestyle for their growing family, a preemptive job offer for Edwin, or both. They had already endured a serious health scare in the summer of 2016, a nationwide polio epidemic that infected thousands and killed more than two thousand in New York City, mostly in Brooklyn. (It was likely during this epidemic that my grandmother's father Arthur King contracted polio, after which he always walked with a cane.) With my father approaching his fourth birthday and their second child due in mid-October, their current quarters may well have been feeling cramped. It was time to look around—perhaps for a place where their children could have a yard to play in, where the family might even adopt a dog!

In this context, my grandfather either found or was recruited for a job as a Private Secretary at the recently-opened Boston branch of the National Industrial Conference Board. The NICB had been founded in 1916 in New York to help mollify labor unrest following the Triangle Shirtwaist Factory fire in 1911. The organization turned out to be as liberal as pro-management organizations ever get. During the time my grandfather worked there, the NICB conducted and published groundbreaking research on worker's compensation and the 8-hour workday and developed the Cost of Living Index. (Much later, Alan Greenspan began his career at the NICB.) Today it might be called a think-tank, and it may have been my grandfather's introduction to the publishing business. This new job would explain their choice of New England, a place to which to the best of my knowledge neither of my grandparents had any previous connection.

How did my grandparents and their children manage to survive the Great Influenza? Did they catch a milder version during the initial Spring wave, when they were in Brooklyn, that protected them from the deadlier version that came roaring back in the Fall? Did one or more of them contract the flu that raged around them shortly after they arrived in Massachusetts—but recover? Was it pure luck?

Despite my grandmother's proclivity for journaling and letter-writing, I have yet to find a word in the family archives that might shed some light on this issue. Many others have likewise remarked on how little mention the Great Influenza has received, in either the history books or in literature, either contemporaneously or later. Only recently, in the wake of several other pandemics or would-be pandemics, including the Covid-19 crisis, has this lost pandemic begun to receive the scientific and historic attention it deserves.

MY FATHER'S BRUSH WITH LITERARY HISTORY

My father described his parents, especially his mother, as "pursuers of causes." I picture their household as resembling the Alcotts', with a succession of visits by artists and intellectuals—in some instances people who for political or personal reasons might not have been so warmly welcomed elsewhere.

During their years in Massachusetts, because of my grandmother's profession as an elocutionist, and because Wellesley College was the center of a yeasty intellectual community, my father and his two younger brothers were exposed to a number of minor and minor-major literary lights of the era. In my oral history interview with my father, he singled out two in particular whom he remembered meeting as a boy.

Katharine Lee Bates

Although Katharine Lee Bates is best known as the author of "America the Beautiful," some of her admirers may be surprised to learn about the breadth of her activities and interests, as well as about her long and productive academic career.

Bates was born in Falmouth, Massachusetts in 1859. Her

father, a Congregationalist pastor, died a few weeks after she was born, and she was raised mainly by her mother, a graduate of the Mt Holyoke Seminary for women. She herself attended Wellesley College and subsequently studied at Oxford University. After teaching for several years at the secondary school level, she joined the Wellesley faculty, starting as an instructor and ultimately reaching the rank of full professor of English Literature. In addition to being a popular teacher and mentor, she was a scholar, a prolific poet and novelist, a journalist, and an activist who supported the rights of women, people of color, workers, immigrants, the poor, and oppressed populations in general. She was an advocate for the League of Nations and opposed the American policy of isolationism. She has been credited with helping to found the field of American literature by creating a new college course and writing one of the first textbooks on the subject. She died in 1929.

She was courted seriously by at least two men and appears to have reciprocated their feelings, but in the end she somehow couldn't find a way to remain in the same place with them long enough to cement the relationships. By contrast, she managed to live for 25 years with her beloved friend and fellow scholar Katharine Coman, until Coman's death from breast cancer in 1915, in an arrangement commonly referred to as a "Boston marriage" (or sometimes a "Wellesley marriage"). The exact nature of these close and intense relationships has been much debated, but further speculation about the two Katharines is beyond the scope of this essay.

Bates began writing "America the Beautiful" in 1893, while on a train trip to Colorado ending at Pike's Peak, and published it as a poem two years later. She reworked the words many times, though the basic bones of the piece remained mostly the same. It has been set to music at least 75 times; the one with which we're most familiar was written by Samuel A. Ward in 1882 for his hymn "Materna." I count myself among the many who believe this joyous paean to the multifaceted beauty of our land would make a far more appropriate national anthem than the warlike text of the "Star-Spangled Banner," penned by a slave-owning lawyer and set to the almost unsingable tune of a rowdy English drinking song.

According to my father, Bates was one of my grandmother's elocution students. Since she retired from Wellesley in 1925 at the age of 66, after decades of lecturing, one might well wonder why she would want or need elocution lessons. In fact, Katharine did study elocution in 1885 (before my grandmother was born!) when she was preparing to teach at Wellesley, but Melinda Ponder, author of a fascinating biography of Bates published in 2017, told me she had never come across any indication in Bates's notes or diaries of her ever having done so again. We do know, however, that Bates "abominated" lecturing and all her life considered herself to be a poor public speaker.

Two possibilities, then: 1) She may have anticipated (correctly) being in great demand as a lecturer to wider and more varied audiences after retirement and wished to hone her public speaking skills further. 2) She was part of my grandparents' social circle and visited their home, but my father, only a boy at the time, was mistaken about her having studied elocution with his mother.

Mary Antin

Another visitor to the Stodola home was Mary Antin, an immigrant writer and activist best known for her 1912 autobiography *The Promised Land.*

Antin was born in 1881 to a Jewish family living in present-day Belarus. Her father emigrated to Boston in 1891; his wife and six children joined him three years later. Mary completed her elementary school education and attended the Girls' Latin School, now the Boston Latin Academy.

In 1901, she married Amadeus William Grabaum, a German geologist at Columbia University, and moved to New York, where she attended Barnard and Columbia Teachers' College.

Among her new friends in New York was Josephine Lazarus, sister of the poet Emma Lazarus, author of the sonnet "The New

Colossus," once mounted on the Statue of Liberty. (Its most famous lines: "Give me your tired, your poor,/ your huddled masses yearning to breathe free,/ the wretched refuse of your teeming shore./ Send these, the homeless, tempest-tossed to me.") Josephine Lazarus urged Antin to write her autobiography; after Lazarus' death in 1910, she pushed forward with the project as a tribute to her friend. In *The Promised Land,* she gave an account of the treatment of Jews in Czarist Russia and described her own education and assimilation in her adopted homeland. No other narrative has improved upon the discernment and authenticity with which she captured the Jewish immigrant experience.

In 1914, she published her last full-length work, *They Who Knock at our Gates,* protesting the movement to restrict immigration. The book was well-received but failed to achieve the popularity of *The Promised Land.*

Antin must have been near 40, and at a very sad moment in her life, when she encountered the young King Stodola. During World War I, while she campaigned on behalf of the Allies, her husband threw his support to his native Germany. The resulting rift led to their separation. At the end of the war, her husband was forced to leave Columbia and went to China to rebuild his career—in fact, he is known as the "father of Chinese geology." Although they continued to correspond, Antin and her husband never saw each other again. She never fully recovered from this devastating blow.

After the war, with scant hope of ever being reunited with her husband, she moved from New York to the Boston area, where her family still resided, and this is when she was presumably drawn into my grandparents' orbit.

Antin died of cancer in 1949.

※ ※ ※

Like so much in the oral history interview, I wish I'd followed up on these passing allusions and elicited more information.

How much did Bates and Antin actually interact with him? Did they have any lasting influence on his intellectual life? But even in the absence of answers, I can't help wondering if my father's deep love of language, his encouragement of dinnertime debates, and his favorite saying, "It ain't necessarily so" might not be rooted in the atmosphere of intellectual ferment my grandparents fostered in their home life, and perhaps even in his encounters with these feisty women who devoted their literary talents to questioning the status quo and advocating for a better life for everyone.

HAM RADIO—WHEN ALL ELSE FAILS

THE EVENING WORLD RADIO SECTION, SATURDAY, MARCH 9, 1929

High School Radio Clubs Active in New York

Brooklyn Technical Now Rebuilding Its Fifty-Watt Transmitter

*no caption, but I'm 100% sure that's my
dad on the front row, far left*

My father's "ham shack" adjoined the bedroom I shared with my sister Leslie when we were very young, so the soft beepity-beep of Morse code was as cozy and familiar to us as the sound of the furnace turning itself on and off or the hum of the fridge. I've seen walk-in closets larger than that tiny room, but it had a window, and it was big enough for his rig, his telegraph key, and his collection of QSL cards sent by fellow hams. It was also big enough for a crib, which had to be squeezed in when Sherry arrived in 1948, and the beeping became her lullaby.

Perhaps this is why I've never minded machines that beep at me.

* * *

Amateur radio had many progenitors—pioneers without whose inventions and discoveries it could not exist—not only iconic names like Maxwell, Hertz, Marconi, but also mavericks like Samuel Morse, an American portrait painter who in a moment of inspiration brilliantly stumbled into the concept of telegraphy—the original digital code.

But in the distinctive form we know today, ham radio dates back to the first decade of the twentieth century. It was then that the relevant equipment and components became commercially available (1904), the first magazine aimed at encouraging amateur radio and "home-brew" equipment, *Modern Electrics*, was launched (1908), the first amateur radio club, The Junior Wireless Club, Limited, of New York City, was formed (1909), and call letters or call signs came into use (1909). It was also then that the curious term "ham" entered the vocabulary—probably as a pejorative, as in "ham-fisted amateur", but it was eventually embraced by skilled operators who knew that the correct opposite of "amateur" was not "professional" but "commercial"; hams were doing it for love, not money.

Note that although in those early days (and even in my own childhood) Morse code was almost synonymous with ham radio, voice communication was also possible almost from the beginning.

In 1910 a bill introduced in the U.S. Senate to prohibit "amateur experimenting" was roundly defeated thanks to the lobbying efforts of the Junior Wireless Club. A similar effort in 1919 to limit wireless communication to the Navy and exclude amateurs altogether was also quashed through the efforts of the American Radio Relay League (ARRL). This decade also saw the introduction of licensing (1912), frequency restrictions and operating instructions (driven in part by the shocking sinking of the Titanic in 1912), the regularization of call signs (1913), and the proliferation of ham jargon (ragchew [extended chat], ham shack [radio room], and 73 [best wishes], to list just a few) and the Q signs (QRZ [who are you?], QSO [did you hear me?], and dozens of other abbreviations that facilitated rapid coding). QSL postcards confirming a contact, often quite creative and artistic, were first used in 1916 and became more standardized in content (sending and receiving call signs, date, time, frequency, mode of transmission, and signal report) in 1919.

All amateur radio activity was forbidden during World War I but sprang back when the War ended. By 1920 there were around 6,000 licensed hams including a handful of women (officially dubbed "YL"—Young Ladies—by the ARRL in 1920; yes, really). In 1923, following assignment to the "useless" higher (short-wave) frequencies by the government, amateur radio operators accomplished the "impossible"—dependable wireless communication between the U.S. and Europe using relatively low power. "DX"—hamspeak for sending messages over long distances—had truly been achieved.

* * *

Following a venerable ham tradition, my father was taken under wing as a young boy by an "Elmer," a neighborhood hobbyist who introduced him to amateur radio and showed him the ropes. He was licensed as an Amateur Class radio operator in 1929, when he was fourteen years old. His call sign was W2AXO, in accordance with an elaborate lettering and numbering system indicating that he resided in New York. My mother always claimed, though I cannot prove it, that he was the youngest person to be licensed at the time. Of course, records are made to be broken, and many younger hams have been licensed since then, some as young as five years old.

To earn a license in those days, a candidate had to complete an essay-type question and demonstrate his or her (mostly his) ability to tap out at least ten words per minute before a Radio Inspector at a field office of the Department of Commerce. Radio clubs like the one my father belonged to at Brooklyn Tech were formed in high schools not only to provide members with the opportunity to practice their Morse code skills and prepare for the licensing exam but also to learn to build and repair the equipment they used. These clubs with their obvious geek appeal were a natural nursery for future engineers, and it's hardly surprising that at least four of the five members of the core Project Diana team were hams.

Any opportunities my father and his colleagues might have had for pursuing or sharing this interest during their years with the Signal Corps, however, were cut off a month after the attack on Pearl Harbor, when as in World War I, all ham radio licenses were suspended to prevent spying. (An exception was made for VHF operations on 112 MHz by around 250 hams who were part of the War Emergency Radio Service.) Because of a clash over postwar frequency allocations—mainly involving radio titans Armstrong (FM Broadcasting) and Sarnoff (RCA/NBC TV), but also threatening access by amateur radio operators (represented by the ARRL)—U.S. hams were not allowed back on the air until November of 1945.

The success of Project Diana in early 1946 captured the imaginations of hams everywhere. Project Diana is still revered in ham radio circles as the *locus classicus* of a popular form of amateur radio communication known as Earth-Moon-Earth. EME involves the use of the moon to relay radio signals from one station on earth to another—the main difference, conceptually trivial, being that the object of Project Diana was to return the outgoing signal to its point of origin. By the end of the 1950s, military use of the moon, a direct outgrowth of Project Diana, was discontinued in favor of the newly-developed artificial satellite technology, and EME became available for use by amateurs.

The amateur radio community has participated in and benefited from satellite technology as well, in the form of satellites built by and for the amateur radio community, starting with OSCAR I (Orbiting Satellite Carrying Amateur Radio) in 1961 and continued by AMSAT, an umbrella designation for amateur radio satellite organizations all over the world. AMSAT launched its first satellite in 1970; its most recent satellite as of this writing was put into orbit in 2018.

* * *

When the Stodolas moved from New Jersey to Long Island in 1956, my father set up his ham equipment in his den and erected an elaborate wooden structure in the back yard, dubbed the "Eiffel tower" by my brother Bob. I was too preoccupied with navigating my own adolescence during these years to pay close attention to what went on in the den, though I remember overhearing occasional boring (to me) conversations with other hams comparing their respective Memorial Day parades or Fourth of July fireworks displays. My father also added mobile capability during these years, and Bob recalls endless hours of riding in the car listening to his CQ calls. And although my father seldom used Morse code anymore, he always kept a telegraph key with his gear "just in case."

In the 1970s, citizens band radio started to fill the niche that had been occupied by ham radio in my father's life, and call sign "W2AXO mobile" was joined if not replaced by his CB handle "Road King." By then he was living in Florida, and CB was in its heyday. Because CB had no technical licensing requirements, it offered a larger if less selective field of prospective contacts and a built-in conversation topic of mutual interest to nearby drivers (though as Bob remarked, it's impossible to imagine our dad saying, "Hey, good buddy, there's a smoky near exit 5..."). My father had also started using a computer by this time, and I don't think he ever again set up a ham radio station in his home.

Because of the trend in amateur communications to use voice or digital technologies, but perhaps also with a nudge from the popularity of CB, the Morse code requirement was dropped for technician (entry level) licenses in 1991 and for all classes of licensure in 2007. Although purists lamented the loss of what had once been the hallmark of ham radio, the source (to many of them) of its cachet, the removal of this hurdle gave the hobby a much-needed boost, and applications for new or upgraded licenses increased substantially. (Surprisingly, Morse code has actually increased in popularity rather than declined.)

Nevertheless, Googling the relevance and future of ham radio in an age of digital technology raises some pretty basic questions—starting with whether it even still exists. That one is easy—yes, it definitely still exists. ARRL still registers over 3/4 million licensed hams in the U.S., with over two million more worldwide. Facebook groups on ham and amateur radio have many tens of thousands of members who contribute hundreds of posts per day.

Questions about the future and relevance of ham radio are more problematic. After all, many of the things that today's older hams could do when they were young that seemed so magical at the time—phone calls around the world! free!—can now be done using cell phones, no training or license required.

Does Ham Radio Have a Future?

Despite the healthy census of licensed hams, many are admittedly getting long of tooth, and efforts to recruit younger members via field days, Scout merit badges, or connecting them with "Elmers" (mentors) typically yield only a few new enthusiasts, and even those often fall away after the initial excitement fades.

Is this really a legitimate worry? One ham, after browsing through amateur radio magazines of the 1940s and 1950s, commented on complaints about the lack of youngsters taking up the hobby: "It seems to have been a 'new' problem for a long time." To be sure, it is the kind of activity that will always have a specialized and limited audience—STEM types who also like activities with a hands-on component—but isn't that also part of its appeal?

If the goal of recruitment is to enable the hobby to continue unchanged, frozen in amber, it is almost certainly doomed. If on the other hand the activity is encouraged to evolve, while retaining its central focus on electromagnetic theory and communications—and on engineering new and clever ways to bring them together—then there is less reason for despair: New software-defined radios (SDRs) have enhanced signal processing and detection capabilities for both receivers and transmitters, bringing ham radio fully into the computer age. A pricy transceiver is no longer needed because low-cost "hand-helds," in conjunction with repeaters that pick up weak signals and amplify them to extend their reach, facilitate making contact and joining networks on the go. AMSAT demonstrates that Elon Musk isn't the only one who can send communications satellites into space. In the "right stuff" department, almost all astronauts have ham licenses so that they can educate classroom and museum groups about their mission, or just call home, while circling the globe.

Worries that children will give up hamming when they discover cars and sex strike me as misplaced. Isn't that often the way with childhood interests? I gave up piano lessons when I was fourteen, over my parents' protests, but have returned to music again and again at various times in my life and am grateful for the pleasure it has brought me over the years. And my husband Ovide, who was licensed in his teens only to stop altogether during his career-building years (though he credits the basic electronic knowledge he acquired as a ham with enabling him to set up psychophysiological equipment for his research), resumed and greatly upscaled his ham activity after retirement. (See below for the long version.)

And why limit recruitment to kids? It's a hobby that can be taken up and enjoyed anytime in life. Women in particular, who might have missed out on amateur radio as children because it seemed to be so exclusively a boy-thing, would seem to represent an obvious target for recruitment (preferably with an updated vocabulary in which women hams are hams, not YLs!) and might even bring some fresh new perspectives to the field.

Is Ham Radio Relevant in the Age of Cell Phones and Social Media?

If you're looking for meaningful communication with your fellow humans, try random dialing your cell phone and see what that gets you. Or try having an in-depth conversation on Twitter. And yet hams seeking contacts by calling "CQ" can readily connect with technologically sophisticated people all over the world.

The answer most often offered, however, is the importance of ham radio to civil defense. When faced with a disaster, we rely heavily on a complex network of cell towers that can fail or become overloaded under extreme adverse conditions. Amateur radio, operating independently of this fragile infrastructure, can be counted on—as hams like to say—"when all else fails." This was the rationale for the War Emergency Radio Service (WERS) during World War II, as noted, and for the formation of Radio Amateur Civil Emergency Service (RACES) in 1952, developed within government agencies as a volunteer public service to maintain communications in times of extraordinary need. ARRL also maintains its own civil defense unit, called the Amateur Radio Emergency Service (ARES).

Arguably, the need to produce new generations of people with the skills required to respond in times of crisis is in itself a reason for attracting young people to the hobby. It might just be their chance to be a genuine superhero!

Though "when all else fails" usually refers to a breakdown of infrastructure due to weather, war, earthquake, or other major emergencies, a less dramatic instance of ham radio being the available technology occurred in my own family in 1962 when my sister Leslie spent the summer as an exchange student in Chile. Because telephone calls were prohibitively expensive, my dad suggested she call her family at home by visiting a ham friend of her host family who would then make contact with an American ham who could then phone my parents—awkward, but better than nothing for a homesick teenager.

* * *

During his long quiescence as a ham, Ovide had kept his license active. So when shortly after our retirement the Ann Arbor Hands-On Museum sent out an appeal to all local hams for help in setting up a ham radio station for kids, he expressed his interest, and a seed was planted. Not long after, we went to visit an old friend from my Northport High School days, who just happened to be dating the owner and editor of a popular ham magazine. Our new acquaintance lured Ovide to his ham shack, which Ovide as a licensed ham could legally operate, and sent him home with a backpack full of books on antenna design.

Suddenly, much of the energy and brainpower Ovide had thrown into his scientific research was diverted not only to the study of radio equipment and antenna design, but to electromagnetic theory, near space meteorology, and other such esoterica. Fortunately, his XYL (ex Young Lady—cringe—that is, his wife), having grown up surrounded by receivers and transmitters and steeped in the lore of Project Diana, wasn't fazed by the prospect of investing in an office full of radio equipment or planting a giant hexbeam next to the house.

It turned out to be the perfect hobby, combining his fascination with electromagnetism and radio propagation with his extrovert's love of interesting and wide-ranging conversation (and no, it isn't *all* about radio equipment or whose tower is taller!). He has now talked with hams in 167 countries and counting, including a commercial airline pilot flying 40,000 feet over the North Pole en route to China and a lonely sailor in the middle of the Pacific Ocean on a solo trip around the world. Some of his contacts are brief, but others are extensive and repeated, occasionally blossoming into long-term friendships.

Meanwhile, finding myself immersed once again in all things ham, I did a little research and discovered that no one had claimed my father's callsign since 1992, the year he became a "Silent Key" (SK). So I signed up for HamTestOnline, a self-paced programmed-learning course, spent a couple of weeks intensively cramming, and presented myself to the Witch's Hat Depot in South Lyon, Michigan to qualify for the technician (entry-level) license. The exam itself consisted of 35 multiple choice items drawn from a pool of 400 questions, so there were no surprises; the only thing that varied was the order in which the answer choices were presented, to prevent a test-taker from memorizing only the letters associated with each question.

As of March, 2011, the proud old call sign W2AXO is now back in Stodola hands, waiting for a member of the next generation to take up the mantle. But don't expect to find me by calling CQ; I'm happy to leave on-air conversation to my more gregarious husband.

WHEN KING
MET ELSA

My parents were married on July 22, 1939. It was an event that almost didn't happen. Because it did, and because I love "how we met" stories, the 80th anniversary of their wedding in 1919 provided me with as good an excuse as any to recount theirs.

By way of background, my mother lived in Massachusetts and my father lived in New York City, but they happened to have two mutual friends. One was Grace Boyd, whom my mother had known since her early childhood in Barre, Massachusetts (and who was my source for parts of this story); the other was Mary Baptiste, a college friend from Gloucester, Massachusetts, daughter of a Portuguese fisherman and in Grace's words, "as brash as Elsa was timid." Grace and Mary had both spent time in New York City, and they knew my father through his network of Unitarian young people.

The first time my parents met, my mother had recently graduated from Boston University and was working as a nanny in Beverly Farms, Massachusetts. She was also engaged to be married; heartbroken after the breakup of a long college romance, she had impulsively accepted the proposal of a young man she'd recently met through her friend Mary. One day my father tried to look up Mary in Gloucester during a brief trip to New England and learned that Mary was visiting a friend named Elsa in Beverly Farms. So my father made his way to Beverly Farms, where he found Mary and was introduced to my mother and her fiancé, who was also visiting that day.

And that was that; King returned to his engineering studies at Cooper Union in New York and Elsa resumed preparations for her upcoming wedding. They parted with no expectation of ever meeting again.

Then my mother suffered a devastating blow. Shortly after her wedding, her new husband simply walked away from my mother and the marriage. My mother's only hope for getting her life back on track was a divorce—not only highly stigmatized at that time, but also so protracted and expensive that it threatened to destroy her chances of happiness. Despondent and desperately in need of a change of scene, she moved to the Bronx, finding a job as a mother's helper to earn the money she needed for the divorce.

As my mother's first Christmas in New York approached, Grace sent my father a Christmas card with a note saying, "Elsa Dahart, whom you will remember meeting when you visited Mary Baptiste about a year ago, is now living in the Bronx and doesn't know anyone else in town. Her phone number is...." My father being my father, he gave my mother a call, and although it couldn't yet be said that the rest was history, my mother was gradually drawn into my father's large circle of friends.

An anecdote shared with me by Grace reveals a lot about both of them at that moment. As my mother approached her 25th birthday, her divorce decree finally within reach, she still couldn't help feeling that life might have passed her by. After all, she had always planned to be married and have a child by the time she reached 25, and as she began telling her friends, everything had now gone wrong. One day my father, upon hearing this plaint, said, "Well, now that you have lived your life, what do you plan to do next?" This comment struck everyone as so comical that even my mother laughed.

Although my parents had by now become close friends, my mother was quite shocked when my father proposed marriage. Until that moment she had regarded him as "just" a friend; understandably wary of another such commitment when she was still awaiting the official dissolution of her disastrous first marriage, she took several weeks to think about it. But as Grace put it, "King was so kind and understanding, and made her feel he did truly love her—and they had gotten to know each other so well during a long platonic friendship," she finally agreed to marry him.

At this point, my father had received his engineering degree and accepted an entry-level job with the Army Signal Corps in Washington, DC, assigning radio frequencies to military bases. On July 6, 1939, my mother's divorce decree became final; on July 15, 1939, my parents applied for a marriage license in Washington, DC.

A week later, they were married in the All Souls Unitarian Church in Washington, DC.

❋ ❋ ❋

My father had been raised as an Episcopalian but found his way to the Unitarians at the age of 13. It was the perfect match for him, providing the social and intellectual stimulation he craved in the context of a rational approach to religion.

My mother, who had attended the Methodist Church as a child, converted to Unitarianism when she and my father became a couple and it suited her for about the same reasons it suited my father—though she always thought the Methodists had better hymns.

All Souls Unitarian Church was founded in 1821 and has been located at the corner of Harvard Street NW and 16th Street since 1924. The building's beautiful neoclassical facade was modeled on that of St. Martin's-in-the-Fields at Trafalgar Square in London (where coincidentally Ovide and I spent a delightful evening at a baroque music concert during our 2019 trip to the UK). My parents loved All Souls and for as long as they remained in Washington, DC, they continued to be active in the church. Later, after each of their three daughters was born, they brought us back to All Souls to be christened. (By the time my brother came along several years later, that tradition had evidently run its course.)

It was a happy and successful marriage. Said their friend Grace, "I've always felt that King was very good for Elsa. He was so gregarious, and kept his friendships in such good repair, and he sort of pulled her along with him." Though my mother didn't achieve her goal of becoming a mother by the time she was 25, she happily welcomed me a few days before her 30th birthday and went on to have three more children.

I've said before that my parents were legendary among their friends for how well they got along. I've thought about this a great deal. What was their secret? It was definitely not a case of not caring enough to get angry. Obviously they kept the lines of communication open and practiced mutual respect, patience, and kindness—all those good things—but I also think part of the magic was rooted in their 1950s-style division of labor. For the most part, they each had their own spheres of responsibility, and within those spheres each abided by the decisions of the other.

Sadly, their lives together were cut short by my mother's untimely death in 1965 at the age of 52. Her dreams for their old age and all the wonderful things they would do together after the children were grown never materialized. Possibly because of the aforementioned division of labor, our father was ill-prepared to take over her responsibilities. Our family had lost its rudder.

ON THE MARCONI TRAIL

my dad at the Marconi Wireless Station,
South Wellfleet, Massachusetts

G uglielmo Marconi showed an early interest in and apti-
tude for radio engineering, sending a wireless message
from his bedroom to his mother's garden at the age of
sixteen. Fortunately for him, his parents had both the inclin-
ation and wealth to support his talents. In his early twenties,
failing to generate much interest in his work in his native Italy,
he relocated to England, capitalizing on his Anglo-Irish mother's

connections and money, and also spent much time in America, combing the coasts of both continents for spots suited to transatlantic radio communication, and leaving many wireless telegraph stations in his wake.

Among Marconi's hand-picked sites was the Belmar Marconi receiving station, located on a hilltop on the south bank of the Shark River Basin, at what later became Camp Evans, home of Project Diana and a stone's throw from my childhood home in Shark River Hills. The original buildings were constructed between 1912 and 1914 by the JG White Engineering Corporation for the Marconi Wireless Telegraph Company of America as part of Marconi's "World Encircling Wireless Girdle" project. Weak transatlantic signals received at the Belmar Station were then relayed via a landline connection to a high-power transmitting station in New Brunswick, 32 miles to the northwest.

In 1913 Marconi returned to his native Italy, where he and his wife became part of Rome society and he was made a member of the Italian Senate. During World War II he was placed in charge of the Italian military's radio services. Sadly but perhaps inevitably, Marconi later in life joined the Italian Fascist party and became an active defender of its ideology.

Today we honor him not for his politics but for his work as a visionary who, ahead of his time, dreamed of a connected world and dedicated his life to making his dream come true—an accomplishment for which he shared the Nobel Prize for Physics in 1909.

* * *

Several decades ago my husband (call sign K8EV) and I had to cut short our delightful meanderings in southern England at the Devonshire-Cornwall border in order to keep an appointment with some friends in the Lake District. So we jumped at the chance during a subsequent visit to the UK in 2010 to pick up where we left off and spent the better part of a fortnight exploring Cornwall.

An unlikely (to me) highlight of our tour turned out to be a visit to the Marconi Centre in Poldhu, site of the Poldhu Wireless Station, where Marconi claimed to have transmitted the first east-to-west transatlantic radio message in December, 1901, to his station on Signal Hill, St. John's, Newfoundland. The Poldhu station was built partly on enclosed pastures that remain to this day, making it hard to keep antennas in working order because cows apparently have a taste for coaxial cable.

The original station was designed by John Ambrose Fleming, who invented the ancestor of all vacuum tubes, earning him the sobriquet "father of modern electronics." That station was decommissioned in 1934 and demolished in 1937, but six acres of the grounds were given to the National Trust in 1937 and more land added in 1960. The Marconi Centre, built on the site in 2001, houses a Marconi museum and also provides meeting space for the Poldhu Amateur Radio Club, GB2GM. It took a little persistence to find the site but we ended up spending a fine afternoon perusing the museum's fascinating exhibits and chatting with the local hams.

* * *

Fast forward to 2014, when my husband and I, car-trekking across the Canadian Maritimes, realized we were within striking distance of another Marconi historic site, this one at Table Head in Glace Bay, Nova Scotia, where Marconi transmitted (this time indisputably) the first west-to-east transatlantic message in December, 1902. Again, it took some looking to locate it despite my best efforts at iPhone navigation. Unlike Poldhu, there was no welcoming committee, just an isolated commemorative plaque to mark the start of a cold, windy walk to the edge of a cliff. Nonetheless, we added another notch to our Marconi belt.

* * *

After I thought I'd long since finished going through my father's collection of documents and photographs, my sister Sherry found one last box in her garage that she sent to me to deal with in my role as family archivist. Included among the photos was a small packet of faded images I'd never seen before but nonetheless looked oddly familiar. I quickly realized that they dated to the 1970s and that I was looking at yet another Marconi pilgrimage, this one by my dad (call sign W2AXO), to Wellfleet, Massachusetts. Marconi had selected Cape Cod because of its easterly location and, after scouting a couple of other possible places, settled on an 8-acre parcel in South Wellfleet. In 1903, he persuaded President Theodore Roosevelt to send a message to King Edward VII of England, conveyed in Morse code to the Poldhu Station in Cornwall. Expecting only a confirmation from the Glace Bay Station in Nova Scotia that the message had been relayed to England, they instead received an immediate reply from King Edward himself.

❋ ❋ ❋

I think I'm starting to see a pattern here. Although the bleak, windswept Marconi stations that dot both Atlantic coasts are not most people's idea of vacation resorts, they provide radio buffs with a window into Marconi's mind, a place where they can stop and speculate about why Marconi, surrounded by his surveyors' maps, preferred this promontory to the next outcropping up the coast. Somehow I suspect there will be more Marconi pilgrimages in my future.

It's what hams do.

AN INTERVIEW WITH MY DAD

Dad and Rose in 1976, not long before the oral history interview

On the morning of April 6, 1992, as I was on my way out the door to report for jury duty, my stepmother Rose called from her home in Florida to tell me that my father had died peacefully during the night. My first impulse was to start packing, but Rose thought it would be better to postpone the memorial service until a time when we could all make arrangements to attend, and I conceded she had a point. So there I stood, with my coat on, feeling slightly stunned. I wasn't sure what to do; I wasn't needed in Florida, so I just went and did my jury duty. Fortunately I was not selected to serve on a jury; I just sat in the waiting room all morning, tears streaming down my face.

Later in the week, I mentioned to my sister Leslie that it felt very odd to be going about my business as usual, when what I really wanted to do was shout, "Hey, World, stop! Something's missing! Someone's gone!" Leslie replied that her response to this dilemma had been to listen again to the oral history interview I'd conducted with our father in 1979. I thought that was a wise suggestion, and over the next few days my husband and I spent our dinner hour doing the same thing.

I expected hearing my father's voice to be therapeutic because it would give me an opportunity to reflect on him and his life as I felt the need to do in the immediate aftermath of his death. It did that, but it also turned out to be therapeutic in a way I didn't anticipate: It took me back beyond those last terrible two or three years of watching the tragedy of a brilliant man with twenty patents to his name forget how to use his computer and unable to work his answering machine, and brought back a larger perspective on his life as a whole.

* * *

I caught my father at perhaps the ideal moment for our interview. He had had major surgery a few weeks earlier; he was now nearly recovered but not quite ready to return to work full-time. He was just about to turn 65 and was considering various retirement schemes; at this crossroads in his life, he was very receptive to my proposal that we use that time to capture and preserve on tape his personal and professional history.

I had recently completed a two-year stint as editor of a grant-funded oral history project, so I not only recognized the importance of seizing the opportunity but also had a pretty good idea of how to go about conducting such an interview. I prepared a well-elaborated "interview schedule," starting with close-ended questions that were easy to answer (what was your mother's name? when were you born? etc.) so that he would be relaxed by the time I got to the tougher and more thought-provoking questions. I took along two portable tape recorders and mikes (to ensure a backup if either failed) and plenty of pre-labeled audio-cassettes. When I arrived at his home, I identified a couple of comfortable chairs in his living room, where over the course of the three days I had set aside for this project, October 2, 3, and 4, we talked for more than five hours, limiting our sessions to a couple of hours apiece to avoid tiring either him or myself.

Any fears I might have had about persuading him to open up proved unfounded. Although he always insisted that he remembered little about his childhood, over the course of the interview I was able to elicit many anecdotes I had never heard and details about his past of which I was only vaguely aware. I think that like me, he had prepared for the interview and spent some time reflecting on what made him the man he was—formative experiences, the importance of friendships and of mentors in his life, his hard-won education at highly selective institutions, his long affiliation with the Unitarians, the development of his career-building skills, and of course his family commitments. In all, we covered the entire sweep of his life, starting with his parents and their roots and ending with a retrospective on his career as he started looking forward to a new life stage. Many of the essays in this volume draw heavily on this material.

In the half-dozen audiocassettes that resulted from this effort, my family has something more precious than anything else he could have left to us.

* * *

A few months after his death, when my stepmother, siblings,

and I mulled over what we could do to commemorate the life of King Stodola at his memorial service, it occurred to me that the oral history interview might have the same healing effect on those in attendance as it had on Leslie and me, that of restoring to us in some measure the man as we knew him. From the several hours of material I'd gathered, I considered what he might want to say to the world at that moment and settled on what he saw as his proudest achievement, his participation in Project Diana.

And so the entire congregation listened, in rapt silence, to the man we all knew before his intellect and his spirit fell prey to the ravages of Alzheimer's, as he recounted the now-familiar story of Jack DeWitt and his dream, the growing conviction that they had the resources to "do the trick," the spine-tingling moment of success, the resolution of their bosses' doubts, the decision to announce the achievement at an Institute of Radio Engineers dinner, and the acclaim that followed. "Philosophically and sociologically," he concluded, "I think it was important because as far as I know, it really was the first time that man had in a measurable way manifested his influence beyond the reaches of the immediate locale of the earth. Indeed, somebody on the moon could have received radio signals that were transmitted from the earth, but this was really the first proof of it.... Other people were working on it, too, and several others succeeded, but we, by circumstance or whatever, were the first."

PART III: MY JERSEY SHORE CHILDHOOD

A. LIFE IN POSTWAR AMERICA

SHARK RIVER HILLS: A MOMENT IN TIME

summer fun with my mom at the beach

Although people have been living in what is now Shark River Hills since Pre-Columbian times, its modern era dates back only to July 8, 1923, the day property in this shiny new resort area went on sale. The Shark River Hills Company had actually purchased and platted the 728-acre tract several years earlier, but initial attempts at development had faltered. Now the Company had placed its renewed hopes in the capable hands of Morrisey & Walker, Realtors. The realtors quickly proceeded to live up to their reputation for aggressive marketing, splashing a large ad in the *Asbury Park Press* on July 7 that proclaimed Shark River Hills to be nothing less than "the LAST HIGH-GRADE DEVELOPMENT NEAR ASBURY." Parcels could be had for as little as $95, or $10 down and $10 per month.

The pitch was primarily aimed at the summer crowd, who were lured by the vision of a vacation paradise and assured that building "YOUR bungalow on YOUR lot" was a better investment than a series of summer rentals. But those who wished to make Shark River Hills their "permanent house" were welcome as well, so long as their money was green.

A driving force behind this sales campaign was Alice "Uppie" Updegraff, the doyenne of Jersey Shore realtors. For years she commuted from her home in Matawan to her job at Morrisey & Walker in Asbury Park, but in 1924 she decided to set an example for her potential customers by buying—not just any house, but the oldest home in Shark River Hills, built in around 1913, a charming bungalow on Riverside Drive to which a garage was later added. The house appeared in many ads for Shark River Hills property. (Later still, the house would be moved to its current location on Glenmere Avenue.)

I knew Uppie from practically the day I was born and have several pictures from a photoshoot on her lawn when I was 4 1/2 months old—me with my parents, me with Uppie's cat—all with her famous house looming in the background, blurred but visible. I suspect Uppie counted my mom and dad as being among a few hundred of her closest friends. More than once I heard the story of her comment when she first saw me as a baby: "Lots of head above the ears—just like her father." I don't give her much credit for phrenology; but for finding a way to admire a baby and flatter the baby's father at the same time, she definitely deserved the Dale Carnegie award for knowing how to win friends and influence new parents.

I remember Uppie as a tiny, wizened old lady though at the time she was probably about the same age as I am now. Living with her in the bungalow were her willowy daughter Virginia—also a real estate agent—and Virginia's husband, the splendidly mustachioed Alphonse Tonietti.

I always found Alphonse a bit of an enigma, but with the help of my friend Joyce, who has a better memory than I do for these details, and my cousin Alan, who is a better scholar than I am, I've pieced together most of his story. Alphonse was born in 1896 in Basrah (then part of the Ottoman Empire), attended the American College in Beirut, Lebanon, and arrived in the U.S. in 1921, where he graduated from the Columbia University School of Journalism. He went on to have what seems to have been a fairly distinguished career, working as editor of the *Literary Digest Magazine* and serving on the staffs of the *New York World-Telegram* and the *New York Sun*.

Then, in the mid-1930s, Alphonse achieved a measure of possibly unwanted fame when he was fired by the Italian language newspaper *Il Progresso*, followed by a lawsuit that was widely covered by the Communist and liberal press. Amazingly, he won his case and was reinstated with seven weeks' back pay, the Court finding that he had been discharged not because (as *Il Progresso* claimed) he didn't write in Italian—that had never been a requirement of his position as editor of the American page—but because of his activities as Chair of *Il Progresso's* chapter of the New York Newspaper Guild.

By the time I knew him, he had moved on to become the owner and operator of the Holy Land Art Company in New York City, a manufacturer of religious artifacts and fabrics. He died in his office on Murray Street in June of 1958, at the age of 61. Virginia Tonietti lived on for another thirty years; she never remarried.

Two details of more than passing interest are missing here—why Alphonse abandoned his journalism career in favor of selling religious objects; and even more intriguing, how on earth he managed to meet an all-American girl from the tiny hamlet of Shark River Hills, New Jersey and persuade her, in October of 1931, to become his bride.

* * *

When my father left the Signal Corps at Camp Evans in 1947 to take a position with Reeves Instrument Corporation in Manhattan, it likely came with a substantial pay raise, and that is probably when my parents decided to give up their rental unit on Clinton Place and go house-hunting. It goes without saying that Uppie and Virginia were close at hand to shepherd them through the process. The house they chose was a modest bungalow on Pinewood Drive. It seemed immense to me at the time, although it is currently listed on Zillow as being 1,879 square feet, a figure that includes two large porches that have been enclosed since we lived there, plus a substantial addition. So—a not-so-big house back then. According to Zillow it was built in 1908.

But wait! Wasn't Uppie's house, built in around 1913, supposed to be the oldest house in Shark River Hills? Surely Zillow had gotten the date of the Stodola home wrong.

When I visited Shark River Hills in 2017, I had a chance to chat with the current owner of the house, who assured me that 1908 was the date shown on the deed. When I observed that the date seemed too early by several years, he told me that the daughter of the previous owner, the one who had bought the home from my parents, had informed him that the house was originally built not as a residence but as a retreat for Roman Catholic priests, on land purchased by the sister of one of the priests. In fact, he said, the house featured two small prayer rooms, one in the basement, the other the second-floor enclosure barely larger than a closet that my father had used as his ham shack!

So far my efforts to track down the previous owner's daughter or to find any corroboration of the house's supposed history as a Catholic retreat have failed.

* * *

Although Uppie and others made it their full-time home, Shark River Hills remained largely a resort community for quite a while. Relics of this history were all around when I was a child; across the street from us on Hampton Court was a log cabin (still there!) that was shuttered most of the year, and our closest neighbor was a tiny cottage only occupied during the summer months. Shark River Hills even had its own boardwalk, running from the Club House to the Tucker's Point Bridge, which had doubled the number of routes between Neptune City and Shark River Hills when it was completed in 1923.

But the little subdivision was on the verge of a sea change. As our friend and neighbor Mary Jane Evers put it, Shark River Hills in the early 1940s was "a summer development, [with] homes not equipped for year-round living, and about three macadamed roads which the Army had done to get to their own properties in the Hills and to allow the personnel to get to work." My parents' circle of friends were recent arrivals, just as we were, and they were all full-timers.

As usual, Mary Jane had it just right, even if there's no precise demographic term for an old-timey summer resort proceeding peaceably along its journey toward year-round development when it suddenly finds itself host to an Army base and under pressure to make itself user-friendly for a large infusion of military and associated civilian personnel. My father was part of that Camp Evans infusion, and so it happened that this oddly unique moment in the history of Shark River Hills formed the backdrop for my childhood.

The "great hurricane of 1944" demolished the old boardwalk, and no one bothered to rebuild it. The era of Shark River Hills as a summer resort had come to an end.

THE DARK SIDE OF JERSEY SHORE HISTORY

I f it weren't for "Birth of a Nation," it might never have happened, at least according to some historians. The release of D. W. Griffiths's inflammatory silent film in 1915 immediately triggered a revival of the Ku Klux Klan and provided it with fictional historical roots by implying it was simply a continuation of the original Reconstruction Era KKK. The new incarnation sought to don a cloak of quasi-respectability by adopting the practices of established fraternal organizations, complete with passwords and secret handshakes. Like its predecessor, it advocated white supremacy, but it also cast a broader net, including Catholics and Jews as targets of its animosity. This helped it gain a foothold in rural and urban areas alike, and by the 1920s it had spread to broad swaths of the country.

New Jersey was no exception. Here as elsewhere, many White Protestant communities felt threatened not only by African Americans moving north but also by the waves of mostly Catholic and Jewish immigrants from Eastern, Central, and Southern Europe, all seeking freedom and competing for economic opportunity. In response, KKK activity erupted in various spots around New Jersey—including very close to home.

From 1925 to around 1932, Belmar Station, later the site of Camp Evans (known, ironically, for its policy of employing top scientists regardless of race or ethnicity), was owned by the Monmouth Pleasure Club Association, a front for the Ku Klux Klan.

The Belmar Station property was a lavish spread that included the Marconi Hotel, a power plant, a number of wooden outbuildings, and around 90 acres of land, later enlarged through the purchase of additional land. It was listed in contemporary documents as the Klan's New Jersey State Headquarters and became the epicenter of Klan activity in New Jersey. For several summers the organization sponsored a circus behind the Marconi Hotel. It also maintained one and possibly several giant crosses that were lighted up on summer evenings, a spectacular but terrifying sight for picnickers across the river on the beach.

Apparently the original hope was to create a summer retreat by selling subdivided lots to Klan members. This plan was thwarted by prolonged litigation between factions of the membership and ultimately by the Depression.

In November of 1941, the property was acquired by the Army and just days later Pearl Harbor sent their operations into high gear. If my parents were even aware of the area's history as a KKK stronghold, it certainly wasn't a part of *their* Neptune, nor of mine. They and their friends were newcomers themselves, and many were Jewish, Catholic, or African American. Besides, they had a war to fight and children to rear.

An odd sidelight: The Monmouth Pleasure Club had hosted meetings attended by national KKK officials, whose "grand wizard" was Hiram K. Evans of Illinois. In his honor, the Marconi site became known as the "Evans Encampment." Years later, the site of the Signal Corps Radar Laboratory, in a solemn ceremony held on March 31, 1942, was designated Camp Evans in memory of Lt. Col. Paul W. Evans. A mere coincidence?

SPAM

I was only two years old when World War II came to an end so have no recollection of the rationing of gas, meat, sugar, or any of the other staples that were distributed to citizens only in controlled amounts because supplies were needed to support the War effort. Nor do I remember ever hearing anyone talk about it, nostalgically or otherwise. All rationing ended in 1946, and with a great sigh of relief Americans began consuming with a passion that has only grown since then. Rationing, it seems, was a subject best forgotten.

I do, however, remember two by-products of rationing—items to make do with when you can't have the real thing—that persisted into subsequent years. The first is the zinc-coated steel penny minted in 1943 to conserve copper, which was needed for ammunition and other military equipment. These pennies were lighter in weight than their copper counterparts, and unlike any other American coin ever, they were magnetic. They continued to turn up like—well, like a bad penny—until the 1960s, when the Mint finally succeeded in rounding up and destroying most of them. They can now be purchased for a few dollars apiece; only a handful of 1943 copper pennies and 1944 steel pennies, both struck by accident, have any real value.

The second is Spam.

Spam is a canned cooked meat product with an almost indefinite shelf life (claims the manufacturer; or two to five years if you go by its "best by" date), made with pork, ham, salt, water, modified potato starch, sugar, and sodium nitrate. Natural gelatin is formed during cooking and leaves a jelly-like coating in the can. One 3.5 oz serving is laden with enough salt (57% of your recommended daily intake), fat (41%), and preservatives to make a nutritionist cringe. On the other hand, it has only six ingredients (not counting water), none of them unpronounceable, bringing it close to complying with the "five-ingredient rule"— Michael Pollan's famous criterion for "real food."

Born of the Depression, Spam was introduced in 1937 by the Hormel Foods Corporation of Austin, Minnesota. The name was invented by the brother of a Hormel executive, who won a $100 prize for his submission. Hormel claims that it means something, but exactly what is a closely-guarded secret (perhaps forgotten by now). Speculations include that it is an abbreviation or acronym for "SPiced hAM", "SPAre Meat", "Shoulders of Pork And Ham", "Specially Processed American Meat", "Specially Processed Army Meat", and a large number of less printable variants.

Although Spam had already been around for a few years, it was during the war that it became a staple of the American diet —because unlike most other meat products it was not rationed. Because of the difficulty of getting fresh meat to the Front, it was also ubiquitous in the military, serving as the WWII version of MRE. The Army found other uses for Spam as well—to grease guns, for example—and the cans were used for scrap metal. Consequently, the majority of Americans old enough to qualify for Medicare probably remember eating Spam. (And like me, they probably remember the distinctive can that opened with a key.)

My mother, true to her New England roots, served it pan-fried with baked beans. I don't know that I ever clamored for it, but I really don't remember it as a penance, either.

* * *

Having assumed that Spam had long since been sent to the abattoirs of history, I was somewhat surprised to notice a few years ago that it could still be found on supermarket shelves (now with a convenient pop-top instead of a key) and on amazon.com, where Spam Classic boasts a 4.5-star rating. (There are also a dozen and a half Spam spinoffs, including Spam Jalapeño, Spam Chorizo, and Spam with Hickory Smoke, not to mention Spam singles and Spam spread.) In fact, it has never gone away at all. For some it's a cheap source of protein; for others it's a comfort food. By 2003, it was sold in 41 countries and trademarked in more than 100 countries (excluding most of the Moslem world). Another Spam milestone came in 2007, when the seven billionth can of Spam was sold. It is still manufactured in Austin, Minnesota, where Hormel maintains a restaurant (Johnny's SPA-Marama) with a full menu of Spam dishes and a Spam Museum. Spam is also manufactured in Fremont, Nebraska.

Over the years, Spam has taken its place in popular culture, turning up thinly disguised in *Mr. Blandings Builds His Dream House* (in which Cary Grant, who works for an ad agency, is assigned to the "Wham" account); in "Weird Al" Yankovic's take-off on the R.E.M. tune "Stand"; as well as in many more obscure locations.

But by far the best-known parody of Spam is an iconic 1970 *Monty Python* skit in which the server in a cafe offers "her" customers "Spam, Spam, Spam, egg and Spam; Spam, Spam, Spam, Spam, Spam, baked beans, Spam, Spam, Spam, or Lobster Thermidor aux crevettes with a Mornay sauce served in a Provencale manner with shallots and aubergines garnished with truffle pate, brandy, and with a fried egg on top and Spam." Meanwhile, a group of Vikings who also happen to be eating in the cafe (where else would a Spam-craving Viking hang his horned helmet?) periodically bursts into song in exuberant praise of "Spam, Spam, Spam, Spam, Spammity Spam, Wonderful Spam." It would probably be safe to say that the *Monty Python* sketch is more famous than Spam itself, and that except for the oldest among us, many would never otherwise have heard of Spam. The source for the use of the term "spam" to refer to unsolicited and unwanted email is almost certainly *Monty Python* and not Hormel.

Is there a connection between spam and Spam, beyond the fact that both are something you might prefer not to appear on your plate? The answer appears to be no. Since the information technology world was (and is) populated by *Monty Python* fans, the simultaneous appearance of the Spam skit about mystery meat and the felt need for a term for mystery email seems to have been pure serendipity.

Will Spam ever make a comeback? Can its humble image be polished? After I learned that some chichi Manhattan restaurants added dishes with tobacco as an ingredient to their dessert menus when indoor smoking was banned, nothing would surprise me. Accordingly, I Googled "spam served in fancy restaurants" and sure enough, up (among several entries) popped a 2015 blog post entitled "SPAMalot! Look at How These Trendy Chefs Are Using Spam," featuring such multi-ethnic entrees as Spam Sliders, Spam Sushi Dog, and (in a pairing that startled even jaded diners-out) Foie Gras and Spam Loco Moco.

Shades of *Monty Python*.

GROWING UP IN AN AMUSEMENT PARK

my mom with her in-laws

L et's start with the boardwalk.

The world's first boardwalk opened in Atlantic City on June 26, 1870. Today the Jersey shore boasts more board-walks than anywhere else in the U.S., with nearly every coastal town having one, some extending from one town to the next. If you grew up on the Jersey Shore, the boardwalk was as much a feature of the natural world as a mountain or a lake might be in some other place.

Who doesn't love a boardwalk? It was a place where people

could come to see and be seen, to admire each other's finery and the incomparable view of the wild Atlantic. It was a great equalizer—free, so rich and poor alike could enjoy its blandishments. It was a place to stroll and relax, a place guaranteed to please your houseguests, be they visiting dignitaries or simply friends and relatives. The above photo, which dates to around 1940, shows my mother on the left entertaining her inlaws—my Uncle Quentin seated next to her, Uncle Sid leaning breezily against the rail, flanked by my grandparents. My father must have been taking the photograph. No dressing down for the weekend here!

Contemporary boardwalks are bustling centers of economic activity, including upscale shopping and fine dining establishments, family and adult entertainment, water parks, beautiful antique arcades, and the like. The boardwalk I recall, however, was a slightly tawdry affair with a sort of carnival atmosphere —though it must have been fairly innocent because we were allowed to go there alone at a relatively young age. I assume the mob violence so graphically depicted on *Boardwalk Empire* ended with the repeal of Prohibition, but if any criminal activity persisted, it flew right over my head. I was more interested in spending my pennies on saltwater taffy than on anything illicit.

What I remember most vividly, however, are the rides (as in, "Mommy, can we please go on the rides? Pretty please?"), and mostly the rides for little kids (remembering that I moved away when I was thirteen). The funhouse was a perennial favorite. Also high on the hit parade (in more ways than one) were the bumper cars; my father took us, which was part of what made the bumper cars so special, and it would be hard to say who enjoyed it more, the children or their dad. Among the most memorable rides for me was the tilt-a-whirl, mostly because a friend and I went early in the season one year, when hardly anyone else was around, and the teenage kid who ran it thought he was giving us a treat by letting us stay on for maybe half an hour. After that, I never wanted to see another tilt-a-whirl in my life. Now, nearly seventy years later, the thought still makes me queasy.

Best of all was the merry-go-round. When we were infants we

maintained a wobbly perch on the stationary animals, held in place by our parents who rode the merry-go-round by our sides. As we grew bigger, we rode alone, and eventually we graduated to the horses and tigers that moved up and down the mounting post in a pattern that was supposed to simulate galloping. The last frontier was the gold ring. Even at twelve, I had to stretch to my limits to reach the ring dispenser. Most of the rings were steel gray, but occasionally a gold ring was dispensed; and your reward for catching the gold ring was—a free ride and a chance to try again for another gold ring. There's a life lesson there, I'm sure.

At the time, of course, I had no idea when I hopped aboard the merry-go-round that I was tapping into a tradition that can be traced back to medieval jousting matches in which riders galloped in circles while tossing balls in the air. By the 17th century, the balls had been replaced by small rings hanging from poles that the riders were supposed to spear—presumably the antecedent of the gold ring. By the mid-19th century the modern platform carousel had been developed (in earlier versions the animals hung from chains and riders would fly out via centrifugal force as the carousel picked up speed) and was a popular feature at fairs. In the U.S. a thriving carousel industry grew up, staffed by skilled craftsmen who created ever more elaborate and lavishly decorated animal figures. Although most of these expert carvers were European immigrants, the animals on the Asbury Park merry-go-round were carved by the American craftsman Charles I. D. Loofa—though a few had to be purchased elsewhere at the last moment to meet a delivery deadline.

My love affair with the merry-go-round puts me in good company. The author Stephen Crane, who grew up in Asbury Park, loved the Palace carousel and courted his girlfriend with nonstop rides in the early 1890s. Asbury Park's other "favorite son," Bruce Springsteen, was filmed riding the carousel in 1987 for the music video "Tunnel of Love"—though alas for carousel fans, that footage ended up on the cutting room floor.

Sadly, the horses have now been put out to pasture and

the three (!) splendid band organs silenced. Palace Amusements closed its doors in November of 1988 and was demolished in 2004, despite being listed in the National Register of Historic Places and despite strong protests by local preservationist organizations. Interestingly, just before the demolition, urban archeologists dug up hundreds of rusted rings; the Palace was still using its ring machine until the day it closed, long after most amusement parks had discontinued the practice in the face of rising costs for rings and insurance.

The carousel was moved in 1990 to an amusement park somewhere in the South. A few years later, when that facility folded, an optimistic New Jersey real estate developer purchased the operating mechanism and returned it to New Jersey in hopes of reinstalling it in a restored Palace building. That hope fizzled when the Palace was demolished, and the remnants of the merry-go-round turned up for auction on eBay. Any attempt to restore the ride to its former glory would require replicas of the figures since most of the originals were long since sold off individually and disappeared into private collections. One more treasure on the junk heap of history....

This photo, taken in July 1955, subsequently appeared in the

Asbury Park Press—an apparent case of "The photographers/ will snap us/ and you'll find that you're/ in the rotogravure." In actual fact, Carolee, Joyce, Susan, and I were recruited to pose for this feature, and I remember much debate about what to wear for the occasion. Naively, we thought the photo shoot was all about us; in retrospect, our job was probably less to show off our Sunday best and more to show off the gorgeous old merry-go-round.

Postscript: Efforts to revitalize the boardwalk culture of the Jersey Shore suffered a huge setback in 2012, with the onslaught of Hurricane Sandy. The Belmar boardwalk was completely demolished, and others suffered extensive damage. Asbury Park fortunately escaped the worst of the storm's fury and was able to reopen for the 2013 season—the only Jersey Shore community to do so—with its newly refurbished boardwalk now linked up to the boardwalk in Ocean Grove.

COLD COMFORT

Making home movies was a popular hobby among proud dads across America, probably with considerable pressure from proud moms. My husband and I both ended up with piles of crumbling, vinegary-smelling reels left behind by our fathers, bearing silent testimony to their offspring starring in dance recitals or feeding the ducks. They have now been digitized within the limits of available technology, and over the years I've been laboriously reducing them to a series of YouTube videos that more or less reflect the original length of each photoshoot, typically three or four minutes.

But this essay isn't about home movies, it's about a particular clip that happens to be one of my faves, of my very young mother sledding with her tiny daughter down the driveway of our home on Clinton Place. It is quite an adorable video, of my mom clinging to her baby as they barrel down the hill, then laboriously pulling me back up, then once again sliding gleefully down.

But this essay isn't even about my mother introducing me to the joys of winter sports, or about the hills that gave Shark River Hills its name, or about baby cuteness. It's about a particular part of my mother's anatomy—namely, her legs. Her bare legs. Which is the first thing any woman would notice upon watching this video—or perhaps second, after remarking on the fact that she was dressed in a skirt.

Brrrrr.

One of my daughters asked me if my mother's legs were really bare. "Those weren't white wool stockings?"

No, I don't recall my mother's ever wearing or even owning wool stockings, white or otherwise. Nylons, you might guess, but in fact she likely lacked whatever small protection nylon stockings might have offered—and therein lies the story I wish to talk about.

Nylon was developed in the 1930s in the lab of Wallace Carothers, a polymer scientist with Dupont, and patented in 1938. Although the Dupont people's vision for their invention—the world's first fully synthetic fiber—extended far beyond hosiery, they cannily decided to start by offering women an affordable and less delicate alternative to silk stockings. The new product was introduced with much hoopla at the New York World's Fair in 1939 and went on sale to the general public in May of 1940. Four million pairs were sold on the first day alone. Nylon became a household word, and "nylons" a synonym for stockings. They were more than just an article of women's underwear, they offered hope that modern technology would lift a Depression-weary nation into prosperity once again.

Barely had nylons become one of life's necessities for the American woman, however, when they were snatched away. In December of 1941 the U.S. entered World War II, and all nylon was diverted to the war effort, used for everything from parachutes to rope to aircraft fuel tanks. The only stockings to be had were bought either before the war or on the black market (so no, my mother would not have been wearing nylons in 1943 or 1944). To give the illusion they were wearing proper nylons, some women painted or penciled "seams" down the backs of their bare legs (probably straighter than it was ever possible to get the real thing!). When stockings were reintroduced after the war, consumer demand outstripped supply, leading to mile-long queues and even "nylon riots," with women getting into fist-fights with one another in the heat of competition. Fortunately, Dupont soon rose to the occasion and ramped up production of the coveted garment.

Late in the 1940s seamless nylons became available, but surprisingly they never entirely caught on. (For some, apparently, the seams were part of the mystique.) Nylons, with or without seams, along with the garter belts that held them up, remained women's wardrobe mainstays until the introduction of pantyhose in 1959—ushering in a trend towards higher hemlines and ultimately the micro-mini skirts that shocked us all a few years later. But that is another story.

Christmas tree bonfire—little girls in their leggings

Little girls did not wear nylons until they reached their teens or at least their 'tweens. For the most part we wore skirts or dresses, under which we wore only our underwear When we went outdoors in the winter, however, there was one small concession to the cold at least for the littlest girls—leggings. By leggings I don't mean either the modified tights that are now called leggings or the leg-warmers worn by dancers and dancer wannabes, I'm talking about thick wooly pants held up with suspenders that were companion garments to our winter coats. These leggings were *only* worn outside, so we put them on and took them off again multiple times per day—going to and from school, for recess, and at lunchtime. During the winter, my kindergarten teacher, Mrs. Pizza (pronounced like the leaning tower, not like the pie), spent much of her day helping the little girls into and out of these cumbersome garments. Synthetics had not yet revolutionized cold-weather gear, at least not for civilians, and we spent much of our outdoor recreation time looking and feeling like the Pillsbury Doughboy—unlike our moms, who simply had to grin and bear it.

RAISED BY THE BOOK

In 1946, when I was three years old, a book that changed my life and the lives of many of my contemporaries was published. Almost overnight, Dr. Benjamin Spock's *Common Sense Book of Baby and Child Care* became the parents' bible, for many years second only on the bestseller list to that other Bible.

"Trust yourself—You know more than you think you do" was its reassuring mantra. Parents had previously been counseled to adhere to a regular schedule of sleep and feeding, even if it meant leaving an infant sobbing for hours. Picking up and comforting babies would only teach them to cry more, argued such experts as Emmett Holt and the behaviorist John Watson. Not only might it delay babies' sleeping through the night, it could also turn them into failed adults, ill-prepared to meet the challenges of the real world. Spock, by contrast, encouraged mothers to rely on their instincts, raising their children not by following a series of rigid rules but by responding to the needs of each child as an individual. What a relief this must have been to mothers, especially those who surreptitiously broke the rules, as I can easily imagine my own mother doing, by soothing their crying infants with forbidden hugs and kisses.

Because Spock lived a long life over which his beliefs and positions evolved, it is difficult not to see the book through the prism of its multiple editions. In addition, his controversial involvement in antiwar activism in the 1960s and early

1970s somehow became entangled in people's minds with his childrearing advice. The conservative preacher Norman Vincent Peale, in an oft-quoted sermon, blamed Spock's "instant gratification, don't let them cry" approach for the violent demonstrations that occurred during that era. More immoderate commentators went even further, demonizing Spock for single-handedly, through some unholy alchemy of his New Left politics and his supposed advocacy of permissive parenting, causing the sexual revolution, selfishness, lust, increases in crime, and the general moral turpitude they believed was destroying American society.

To see the version my mother must have read—the version born with the Baby Boom in 1946 and priced at $.25 so everyone could afford it—I purchased a copy of the first edition on eBay. This volume is addressed primarily not to "parents" but to "mothers"—and White middle-class mothers at that. Gay and lesbian families are not anticipated; even divorced and single parents are treated as special cases, tucked away in an Appendix. Worries about childhood obesity are minimal (he devotes 4 pages to "fat children" and 11 to "thin children"), and the vegan diets he later advocated for all children over age two (after crediting such a diet with giving him a new lease on life) are not part of the picture. Circumcision is not yet on the no-list. Sexist language is baked in, despite a certain amount of defensiveness on the author's part. (He devotes a whole apologetic paragraph to explaining that he's called the baby "him" throughout, despite his insistence that "girl babies are as wonderful as boy babies." "Why can't I call the baby 'her' in at least half the book? I need 'her' to refer to the mother.")

Nonetheless, *The Common Sense Book of Baby and Child Care* was revolutionary. Again, it is difficult to convey how novel and refreshing it must have seemed to his first readers in postwar America, because his advice and the version of child development on which it is based have been so thoroughly incorporated into our childrearing thinking and practice. Spock was a pediatrician, not a developmental psychologist, but he was deeply influenced by Freud (he himself underwent psychoanalysis). Indeed, Spock probably did more to smuggle Freud into the popular culture than Freud's writings could ever have done on their own. (At one point he mentions a little girl who complains to her mother, "But he's so fancy and I'm so plain"—almost certainly the inspiration for Mister Rogers's uncharacteristically graphic song "Everybody's Fancy.") As *Time*'s obituary of Spock in 1998 put it, "Surmising that new parents were not yet ready to hear of their infants' oral, anal, and genital stages, Spock simply advised moms and dads not to get alarmed if a baby sometimes behaved, well, oddly. He had learned from Freud that repression could produce catastrophic adult neuroses. Better, he advised, to "wait things out."

Despite accusations of "permissiveness" (always strongly rejected by Spock himself), his approach can be considered permissive only in contrast with the draconian advice then being offered by Holt and Watson. Dr. Spock expects youngsters to be assigned duties, to put things away, to come to the table when dinner is ready, and to be polite to others. He warns against asking "Do you want to...?" or offering too many reasons when requiring the child to do something. It's okay for a child to make mudpies ("it enriches his spirit"), but not in his Sunday best. Although no advocate of spanking, neither is Dr. Spock uncompromisingly opposed to it, regarding it as "less poisonous than lengthy disapproval." The best description is perhaps the one Spock himself chose for the first edition of his book, "common sense." "Trust yourself," he told young mothers—and they did.

❋ ❋ ❋

So who was Dr. Spock, and how did he come to be tapped for the job of raising a whole new generation of American children?

Benjamin Spock was born in 1903 in New Haven CT, the eldest of six children of a rock-ribbed Republican New Haven Railroad official and his wife. Both parents held firm views about childrearing but it was left to the formidable Mildred Houghton

Spock to carry them out. She followed religiously the regimen of Dr. Holt, with his rigid schedules for feeding, bathing, and elimination. She also believed in the moral and physical virtues of a Spartan lifestyle that included sleeping in an icy bedroom and attending a preschool where lessons were taught outdoors, even in winter.

After preparatory training at Phillips Andover Academy, he enrolled in his hometown college, Yale, for his undergraduate education. Though not a particularly distinguished student, he was a superb athlete and after crewing at Yale became a member of the 1924 Olympic team. Following his graduation in 1925, he entered Yale Medical School but later, over the objections of his family, transferred to Columbia's College of Physicians and Surgeons. It was around then that he fell in love with Jane Cheney, a lively and socially conscious Bryn Mawr student, daughter of a wealthy silk manufacturer from Manchester, Connecticut, and in 1927 they were married. Following an all-too-familiar pattern, Jane worked long hours at Macy's to support her young husband's medical education.

Despite coping with various health problems, including the death of a premature infant, as well as with the onset of the Great Depression, Ben and Jane managed to thrive in the intellectual ferment of New York City, where they were exposed to new ideas and a more diverse society than their rather sheltered lives had previously afforded them. Ben became interested in childhood diseases and in the effects of abuse and neglect. He also discovered a gift for interacting with children. When he graduated from P&S in 1929, he was first in his class. Ben had discovered his métier.

Right after the attack on Pearl Harbor, Ben volunteered for military service but was rejected because of back problems. So the Spocks remained in New York, spending their summers in the Adirondacks. A few years earlier he had rejected a proposal by Doubleday that he write a child care book, saying he didn't know enough. Now, with a little more time to relax than his winter schedule had permitted, he began to think once again about such a book—a book that would embody his ideas on child care, a book aimed not at experts but at the end users, the parents.

In the summer of 1943 Ben began writing, with Jane at his side. Well into the night, after regular work hours, Ben dictated and Jane, pregnant with their second child, transcribed (and often rewrote), sometimes suggesting new avenues to pursue; during the day Jane conducted follow-up research and solicited expert commentary. Eventually Ben was able to join the Navy, but even while he was on active duty, Jane soldiered on with the book, carrying out final negotiations with the publishers, indexing, and hammering out last-minute revisions in the middle of the night via long-distance phone calls with her husband.

The book was an overnight sensation—"only...limited," remarked the publisher, "by an inability to get sufficient paper to supply the demand." Revised editions were issued about every ten years. It was eventually translated into 39 languages. Meanwhile, the Spocks, now a family of four, continued to maintain busy lives. Ben traded his clinical practice for an academic career, accepting many speaking engagements and writing additional books and articles. With a more comfortable income, they spent more time vacationing. Jane gamely though cautiously learned to sail in her sixties. As the Vietnam War dragged on, Ben grew increasingly political, even running as a third-party presidential candidate in 1972.

Then, in 1975, Ben announced to Jane that he wished to begin a "trial separation." He went out and rented an efficiency apartment for himself. A legal separation followed shortly, then divorce, and within a few months he married a woman 41 years his junior. Jane was devastated.

What went wrong? Apparently neither of the two was easy to live with. "Though he was blessed with a great deal of energy and maintained a warm and pleasant public presence," said one Spock biographer, "he held impossibly high standards for himself as well as for those around him." As Jane herself put it, "Ben seems like this outgoing, loving, easygoing person, but he really isn't. He's a stern person." His two sons—Michael, born in 1933, and John, born in 1944—remember him as a cold and remote father, devoid of the warmth and affection he advised his readers to bestow on their children. One told an interviewer his father had always made him feel "judged, criticized, scared, beaten down." "I never kissed them," Spock himself admitted. For her part, Jane spent most of her adult life in and out of therapy. She was dependent on alcohol and Miltown (the tranquilizer immortalized by the Rolling Stones as "Mother's little helper") and suffered from intense mood swings. She also harbored a deep resentment about her husband's and society's failure to recognize the extent of her role in creating the child care book and did not hesitate to berate him in public.

In the end, Benjamin Spock, for whatever reason, decided he no longer wanted to be married to Jane.

The prospects for a divorced 70ish woman were of course much more limited than those for her husband. Jane did her best to make lemonade out of the lemons she'd been handed, advocating and running support groups for divorced older women. By and large, however, she lived out her remaining thirteen years a lonely and embittered woman, institutionalized for six months after breaking down altogether and railing against her ex-husband whenever she had the chance. "If it had been a co-authorship, like it should have been," she told an interviewer, "I would have been asked [to be] on television shows, too, and I would have been asked what I thought about things. I might have been more of a somebody. But I don't think he could stand it, sharing the spotlight."

In the fourth edition of the book, just as their marriage was on the verge of dissolution, Spock finally included a generous though long overdue notice of Jane's participation in the form of a full-page acknowledgment at the front of the book. Was that good enough, or did Jane, as she claimed, and as the publisher at one point suggested, deserve more? Ann Hulbert, in her 2003 history of childrearing advice in America, actually becomes quite incensed on Jane's behalf—though as a reviewer of her book in the *New Yorker* cynically observed, "If we had a nickel for

every twentieth-century author whose wife was an unacknow-
ledged collaborator on his books, we could probably pay for the
war in Iraq."

Nonetheless, although there is no universal standard by
which to judge when transcribing, performing background research, fact-checking, recipe-testing, editing, consulting experts, rewriting, and more cross the blurry line into full-fledged
coauthorship, a case could almost certainly be made for Jane's
claim. Of the two, she was the scholar. Spock did little if any
research on his own and claimed the book "really all came out
of my head." A talented writer and prolific letter-writer, he was
probably responsible for the genial, reassuring tone for which
the book is justly famous, and of course he had the medical creds
to back up his words. But Jane appears to have contributed intellectual content as well as technical and stylistic support. She was
the source of his interest in psychoanalysis, for example, and
persuaded him that personality formation was fairly well established by the age of two.

Given the breadth and depth of Jane's participation, I doubt
that anyone could have quarreled with a joint authorship had
it been proposed—and the official inclusion of a woman's voice
might even have increased its appeal (if indeed that were possible, given the book's astounding popularity). So—let's just say
the decision to go with a sole authorship was a call. Who knows
if Jane's life, and their lives together, might have taken a different
turn had Ben called it differently, either from the beginning or
with the publication of a subsequent edition.

❋ ❋ ❋

I have always felt deep down, whatever the demographers
may say, that I belong to the leading edge of the Baby Boom.
I now realize, however, that I have to temper that conviction a
little, because my poor mother had to struggle through my first
three years without benefit of Dr. Spock. Her first pregnancy had

ended in a miscarriage, and when I was born in 1943 (the same year Ben and Jane started working on their book), she feared she was so old (thirty!) that she might never be lucky enough to have another child. In fact she went on to have three more, but when I first arrived she regarded me as little short of a miracle baby. My infancy is so well documented in text, photos, and home movies that I've sometimes thought it would be great source material for a doctoral dissertation on child-rearing practices of the World War II era.

So I reviewed my mother's *two* baby books on little Cindy for evidence of the childrearing ethos that prevailed before Dr. Spock, and sure enough, there it was: "Cindy started her toilet training on Nov. 19, 1943, her ninth month birthday." As Spock observed, when a baby appears to be toilet-trained that early, "It's the mother who's trained." Readiness was everything, to Spock's way of thinking. There may be no harm in starting so early; on the other hand, if the mother makes excessive performance demands the child may end up rebelling in his second year. Better to put it off for a few more months.

Once *The Common Sense Book of Baby and Child Care* came out, it was the first place my mother turned for answers to all her childrearing questions, so much so that I remember sneaking it off her bookshelf to read the chapters on "From Six to Eleven" and "Puberty Development," just to make sure I stayed out ahead of her. I also remember feeling frustrated and annoyed at her stock dismissal of many issues that seemed desperately important to me (somehow I can't remember what they were) with the words, "She's just going through a phase." It felt to me as though she was denying the validity of my concerns in a way that she would never dream of denying her own. I later realized I was probably fortunate that she adopted this relatively relaxed attitude towards my various notions rather than one of worry or revulsion or anger or any one of a number of responses I might have found even less palatable.

As I prepared to write this essay, I was surprised to discover that the man who shepherded me through most of my childhood stages even had a few words of wisdom about my current life stage. In his autobiography *Spock on Spock*, he described his attitude towards aging as "delay and deny"—an approach I too have found useful.

DR. HELEN JONES

I n the mid-1970s, I worked for two years as the Director of a grant-funded Oral History Project on Women in Medicine at the Medical College of Pennsylvania in Philadelphia (formerly the Women's Medical College of Pennsylvania and the last medical school in the country to go co-ed; currently part of the Drexel University College of Medicine). In that capacity, and as co-editor of a book that included several oral history interviews resulting from that project, I learned quite a bit about the history of women in medicine in the United States and the obstacles these pioneers encountered in their attempts to achieve their career goals. I am, you might even say, something of an expert.

So I felt more than a little sheepish when it only recently crossed my mind that my childhood pediatrician, Dr. Helen E. Jones, was a part of that history; and that even being her patient, at a time when only a tiny fraction of physicians were women, was an unlikely feature of my childhood. I immediately recognized this topic as a potential essay. Almost as immediately, I recognized that despite my vivid visceral memories of Dr. Jones—her strawberry blonde curls, her small physique, her all-business demeanor, and of course all those injections!—I knew next to nothing about the biographical details of her life or the hardships she must have endured to obtain her credentials and establish a thriving medical practice in a small coastal New Jersey town.

Having gotten as far as I could with Google and the archives of the *Asbury Park Press*, I consulted two main additional sources. One was my friend Sandra Chaff, Archivist-Director of the MCP Archives and Special Collections on Women and Medicine at the time I worked on the oral history project, who was fierce about finding and preserving documents, photographs, and artifacts relating to the history of women physicians. The other was a Facebook Shark River Hills nostalgia site, several of whose members were also patients of Dr. Jones and were generous in sharing their recollections.

I am now happy to report that with a little help from my friends (thanks, guys!), I have cobbled together at least an essay's worth of information about the elusive Dr. Jones.

* * *

Helen Elizabeth Jones was born on October 15, 1918 in Scranton, PA, the middle of three children of Horace I. Jones and Norma A. Johns Jones. When Helen was three, the family moved to Asbury Park, where her father taught at Asbury Park High School for many years. Her mother was a musician and teacher who ran her own private school.

After her graduation from Asbury Park High School, wishing to see more of the country, she enrolled in premedical studies at the University of Michigan. She also joined the Psychology and German clubs during her days in Ann Arbor and pursued music and dramatics as well. During her summers at the Jersey Shore she worked as an assistant at a first aid station at the beach.

In September of 1940 she enrolled in Temple University School of Medicine—one of around ten women in a class of 104. "Gone are the carefree college days," intoned Dr. John B. Roxby in his welcoming speech. "You have chosen a great but difficult path, and the travail which lies ahead is of a quantity which may defy your present imagination." Along with the demands described by Dr. Roxby, additional clouds hung on the horizon.

The prospect of American entry into World War II loomed larger every day and became a reality at the end of 1941, with the bombing of Pearl Harbor. The Wagner-Murray-Dingell bill of 1943 was as unsuccessful as succeeding socialized medicine proposals but nonetheless caused uneasiness and fear of the unknown.

Helen's graduation photo appears in the December 1943 yearbook—possibly on an accelerated schedule to accommodate a wartime need for medical personnel. The accompanying text notes that she was known as "Jonesy" to her close friends and that her special interests were in pediatrics and obstetrics.

During the latter half of her senior year she did a rotation at the Chestnut Hill Hospital in Philadelphia. Following graduation she interned at the Jersey City Medical Center.

At the end of 1944, she started posting notices in the *Asbury Park Press* that her offices at 617 7th Avenue, located in her home in Asbury Park, would open for business on January 2, 1945, with office hours 2-4 and 7-9 pm, and on Sundays by appointment.

In addition to her regular pediatric practice, over the years she took on additional duties, serving on the staff of the Jersey Shore Medical Center in Neptune for nearly 30 years, as school physician in Ocean Township starting in 1957, and as pediatrician to the Asbury Park Well Baby Clinic for 20 years. She was a member of the American and New Jersey Medical Societies and the New Jersey Chapter of the American Academy of Pediatrics.

In 1974, Helen's brother-in-law accepted a position in California and Helen decided to head west with him and her sister. "I'm worn out physically," she told the *Asbury Park Press* of her nearly 30 years as a pediatrician, adding that her own physician had advised her to give up her private practice. "It was impossible to retire here," she continued. "I did try to cut down, but it didn't work. Children don't get sick according to a schedule. Pediatrics is a day and night, seven days a week profession. Out in California I will be working eight hours a day, five days a week."

Dr. Jagdish Bharara took over her practice in New Jersey, and

Helen sold her house in Deal Park—truly the end of an era.

On May 19, 1974, Helen began her new 40-hours-a-week "day job," which involved working with what were then described as "retarded and emotionally disturbed children" at Camarillo State Hospital. At the time she arrived, Dr. Ivor Lovaas, a pioneer in the application of behavior analytic techniques to autistic children, was at the peak of his career at Camarillo State, so whether or not she worked with him directly, it is impossible that she wasn't touched by his somewhat controversial use of both rewards and punishments to encourage language skills and reduce self-injurious behavior. It would be interesting to know what she thought of his research-oriented approach, in contrast to the treatment model that had guided her own career.

Helen Elizabeth Jones died on December 17, 1993 in El Dorado, Placerville, California, at the age of 75. Four years later Camarillo State Hospital permanently closed its doors.

* * *

One of my fellow denizens of Shark River Hills volunteered: "I remember...the shots in the butt!" Another posted this recollection: "I always went into a panic as soon as I smelled the rubbing alcohol. I broke away from her assistant once and ran around the exam room with the syringe hanging off my butt." Except for the added fillip about the syringe, this funny-but-sad scenario exactly mirrors my own recollection of my mom and Dr. Jones's assistant chasing my little sister around the room as she too made a desperate attempt to escape the dreaded needle.

Although I was too wimpy to engage in such hijinks, I had my own personal nickname for Dr. Jones that still comes more naturally to my tongue than her given name: "Fanny Jones" (yes, for the obvious reason). If you think you detect an undertone of affection in that sobriquet, you would be quite wrong. And yet she never did me any harm; indeed, she probably saved my life on more than one occasion. The only thing I can say in extenuation of my youthful animus is that I remember her as rather severe, so the pain she inflicted was not much tempered by warmth or playfulness.

No one likes having a shot, but I cannot remember this level of needle-phobia in the days when I took my own daughters to the doctor. Did we have to endure more shots in those days? I may be exaggerating but I can't remember a visit to Dr. Jones that didn't include at least one shot. Although the number of vaccine-preventable diseases is larger today than when I was a child, combinations such as diphtheria-pertussis-tetanus didn't come into use until 1948; the shots and their boosters were given separately. Smallpox inoculations continued to be administered (remnants of my scar are still visible) until the late 1960s and beyond, even after the disease had been eradicated worldwide. The long-awaited vaccine that finally ended the terrors of polio became available in 1955, but the orally-administered version didn't make its appearance until 1961. So—for a while, yet another shot.

Non-routine shots were also commonplace—for example, tetanus boosters after injuries, such as I received after a dog bite when I was ten. (Telltale marks where Dr. Jones cauterized the wounds can still be seen on my chin and neck—more battle scars!)

Although the antibacterial properties of the Penicillium fungus had been recognized for decades by the time the Scottish biologist Sir Alexander Fleming succeeded in culturing and concentrating it in 1928, it wasn't until World War II that a form stable enough to be mass-produced for clinical use was developed. Overuse of this miracle drug probably began almost immediately. I became seriously ill with bronchitis and pneumonia when I was in the first grade and Dr. Jones arrived at our home every morning for two whole weeks (yes, Dr. Jones made house calls!), clutching her black doctor's satchel, to give me an injection. More than my fair share—so little wonder, perhaps, that I felt traumatized by my encounters with her!

Alternatively, is it possible that shots actually hurt more back then? Just around the time I began my career as a pediatric patient, glass syringes with interchangeable parts, already a staple of medical practice, began to be mass-produced. Not until the mid-1950s were disposable plastic syringes introduced, eliminating the problem of contamination from improper sterilization; might they also have had a smoother action that reduced the ouch-factor? Or do newer manufacturing techniques produce finer and sharper needles than were available in the 1940s? Does modern medical training focus more on minimizing or distracting from the pain of injections? Just spinning out ideas here, but descriptions by diabetics of changes in insulin injections over time lend credence to the possibility that shots are less painful now than they used to be.

* * *

What led my mother to choose Dr. Jones? I doubt she actively sought out a woman physician, but obviously it didn't put her off, either. My guess is that when my sister was born in late January of 1945, Dr. Jones—having hung out her shingle just three weeks earlier—attended her in the hospital. Perhaps my mother was dissatisfied with her current pediatrician, or perhaps she was attracted by the prospect of joining a new and relatively uncrowded practice. Perhaps she just plain liked Dr. Jones, who, despite her abject failure to win me over, clearly had a way with mothers. Said one grateful mom when Dr. Jones retired, "She takes the time necessary to examine the child and encourage and reassure the parents.... Quite simply, she is a good doctor in the highest sense of the word."

Helen Jones and the choices she made were typical of women physicians of her day. Although women overall were not welcome in the medical profession, pediatrics was considered a "soft" specialty and therefore better suited to women than specialties engaging in heroic treatments of "really sick" patients, or involving surgery. Moreover, women who married were taken less seriously; unlike men, they were supposed to be "married" to their profession.

By the time Helen graduated from medical school, to be sure, these patterns were starting to crack a little. As a result of World War II, women were working outside the home in larger numbers, a trend that undoubtedly helped Helen build her practice. Helen was also fortunate in having as a role model a mother who ran her own business at a high professional level. Nonetheless, Helen played it safe, sticking with a "soft" specialty and remaining single. Failure was probably not an option,

This is not in any way to imply that her choice of pediatrics was simply a matter of expediency. "Medicine is demanding," she said, "but the thing that makes it all worthwhile is seeing a child grow up well and happy…. You become close to the children and they're almost like your own." She maintained a deep conviction of the larger importance of her work: "They will be the heads of state, one way or the other. You look at a little baby and realize that everybody who comes in contact with that child is going to have an influence on the development of his personality and character."

Devoted as she was to her profession, Helen was not completely devoid of outside interests and activities. One of my Facebook "informants," whose cousin worked for many years as Dr. Jones's receptionist, said that Dr. Jones shared her home in Deal Park with her sister and brother-in-law, which afforded some presumably welcome companionship. Her mother's daughter, she was an accomplished musician who played piano and organ. Other hobbies included collecting stamps, with an emphasis on those featuring opera and the history of musical instruments; and caring for her two Yorkshire Terrier stud dogs, one of which took a first one year at the Westminster Dog Show in Madison Square Garden.

Did Helen's busy pediatric practice—stampeding kids and all —add up to a full and rewarding professional life? Did her hobbies enrich her nonworking hours? Did she miss having children of her own, or did she enjoy a little calm and quiet when she returned home in the evening after dealing with other people's children all day? Did she indulge herself in a few precious moments at the piano communing with Chopin, or did she just drop into bed exhausted? Did her second career working with special-needs children and scaling back to a 40-hour workweek in an easier climate give her a new lease on life?

I wonder.

I offer this essay not as a love letter, exactly, but as an expression of gratitude for the medical care I was fortunate to receive as a child and as an apology for the ill will I harbored towards Dr. Jones for no other reason than that she always seemed to be brandishing a needle.

OUT AND ABOUT ON THE JERSEY SHORE

Bill and Barbara Evers by their Willy's Jeep (courtesy of the Evers family)

America's love affair with the automobile was going full throttle in New Jersey when I was a little girl. Construction of the Garden State Parkway began in 1948, followed shortly after by the New Jersey Turnpike, both predating Eisenhower's signing of the Federal-Aid Highway Act of 1956, which created the Interstate System that changed not only the face but also the heart and soul of America.

A subtle but transformative change introduced by this new kind of road was a different idea of the size of our country, of the normal tempo of life, and, on a smaller and more local scale, of the scope of an afternoon's outing. Because going for a ride in the car was exactly my father's idea of the perfect family outing, there were many such Sunday afternoon jaunts in my life. I remember my mother gently touching my father's arm to remind him to slow down—"King, you're going 45 miles an hour." The Burma Shave signs, a series of small road signs on which a clever rhymed advertisement gradually unfolded—much beloved by children including me—presaged their own demise with safety messages like "If you dislike / Big traffic fines / Slow down / Till you / Can read these signs / Burma-Shave," since it soon became impossible to travel slow enough to do so. My father humble-bragged to his friends that his Oldsmobile would pass anything on the road except a gas station.

I suppose part of the fun lay in the optimism and newness of it all. There were no media campaigns to compile and publicize grim nationwide statistics about what happened when things went wrong, or about vehicles that were probably "unsafe at any speed." There were no seat belts, no airbags, none of the many safety features now routinely required by law. Consider, for example, the rumble seat—an upholstered bench that unfolded where the trunk should have been. The last American cars with rumble seats were manufactured in 1938, but used cars with rumble seats must have remained available for at least another decade. Our friends the Everses had one, and we all vied to be one of the two lucky ducks (three, if we wheedled persuasively enough) who got to ride there. It was probably about as safe for a kindergartener as riding unrestrained in the back of a pickup truck, but at that time, who knew?

The Evers family also had a Willy's Jeep. (The Everses had all the good cars!) We Americans loved our Willy's Jeeps. Brainchild of the industrial designer Brooks Stevens, the Willy's Jeep Station Wagon first rolled off the assembly line in 1946, just in time to join American families in their wholesale move to the suburbs. I was pleased to confirm my childhood memory: Despite its appearance, it was actually a faux woody made of painted steel, a design that was both safer and better-suited to mass-production than the wood-bodied passenger wagons of the day. Production continued in the U.S. until 1965, when the Jeep Wagoneer supplanted it in our fickle affections, but continued in Brazil and Argentina for several more years.

Cars of the 1940s were unreliable by today's standards and much more subject to breakdowns—breakdowns that could be repaired not by replacing one electronic component with another but only by someone's getting truly down and dirty. In our household that someone was my father. For him, engineering wasn't just a profession he checked at the Evans Lab door when he came home from work. He was a gadgeteer through and through, and keeping the family chariot in shape for trips to the Cracker Barrel was the perfect busman's holiday for someone who spent his work week tinkering with equipment for sending radio waves on trips to the moon. Among my most vivid weekend memories of my dad are images of his disembodied legs jutting out from under Nancy Nash or Ollie Oldsmobile or whatever our current rolling stock happened to be. It seemed there was always something that needed repair if you looked long enough and hard enough, and he kept his trusty toolbox and flashlight close at hand.

Flat tires were commonplace. One of the many family legends about my father originated with my cousin Tricia, a frequent visitor from Massachusetts, who was astonished to hear my father explaining torque to me as I "helped" him change a tire; I was four years old at the time. This was quite typical of my father, who made little accommodation to our tender years. (My daughters remember their grandfather in much the same way; some things never change.) However unorthodox his pedagogic methods, they must have had some merit, because the tire changing session my cousin witnessed sticks in my mind to this day, complete with my dad's detailed demo of the right way to tighten the bolts, skipping one in between, and the accompanying explanation of torque.

In the summer, a lot of our outings were trips to the beach—with perhaps a stop on the way home for soft ice cream at the new Dairy Queen or Carvel shops. (From my most-embarrassing-moments file: I once threw the unwanted remainder of an ice cream cone out of the Everses' car window, hoping no one would notice I hadn't finished it—and the window turned out to be closed. Oops.)

The journey from Shark River Hills to the ocean beaches, short as it was, included two potential hazards peculiar to the local terrain. One was a large drawbridge that might need to be opened at any time without much warning. Because it could involve a wait of unknown length, beach-bound drivers were reluctant to get caught on the "wrong" side of the bridge, and we watched a number of close calls with bated breath while the guards shooed the cars back in time for the boats to continue through the channel. And heaven help anyone who had to pee when the bridge hove up.

The other notable hazard was a railroad grade crossing, and it was there that I experienced what may have been my childhood's biggest "Rosie moment" (my term for a grave danger of which one is totally unaware, based on the wonderful children's book *Rosie's Walk*, in which a clueless hen is stalked by a hungry fox but remains oblivious to the repeated dangers she narrowly escapes during her morning ramble)—when the double arms of the gate descended with my mother trapped between them, a train bearing down on us and her children squirming in the back seat. I have no idea whether the car stalled or she just miscalculated, but in retrospect, I can only imagine how terrified she must have been. Fortunately, the attendant noticed her plight—the gates were operated manually in those days—and raised the arms in time for us to escape. (Gates that block the entire roadway on both sides of the track are no longer in use for reasons obvious.) What, me worry? I never doubted my mom would get us out of that predicament, just as she always did.

TWO ROUND ATTRACTIONS

T wo important childhood icons are linked in my mind for one reason and one reason only: Both were famous for being round. They carried out "in the round" functions that weren't usually or necessarily done that way (unlike, say, the merry-go-round, another Jersey shore icon—also round, but its roundness was part of its very essence). Otherwise these two institutions have nothing in common, but somehow looking back, when I think of one I always think of the other.

The first was the Rotolactor, a system for milking a large number of cows successively by loading them on a rotating platform. Located in Plainsboro, New Jersey, it was less than an hour's drive from Shark River Hills and only a few minutes from the Unitarian Fellowship we attended sporadically in Princeton. My sisters and I always clamored to stop at the Rotolactor on our way home, and knowing it made Sunday School more palatable to us, my father often complied. Thus, the Stodolas became Rotolactor regulars.

The Rotolactor was first conceived in 1913 by Henry W. Jeffers, an employee of the Walker-Gordon Laboratories in Plainsboro. Development was halted during World War I, but in 1928 the Walker-Gordon dairy was purchased by the Borden Company and development resumed. The Rotolactor. which was put into operation in November, 1930, was sort of a Rube Goldberg contraption including apparatus for bathing, weighing, measuring, and of course the teat-cups. The circular platform was sixty feet in diameter and could hold 50 cows, each of which made the entire circuit in 12.5 minutes. Everything was done mechanically, and an important feature was that the product never came into contact with air and was untouched by human hands. Cleanliness was everything.

The Rotolactor was featured in the Borden's exhibit at the 1939 World's Fair in New York. Borden's had created the cartoon character Elsie the Cow as their mascot in 1936. (Before there was Joe Camel, before there was the Energizer Bunny, there was Elsie.) Discovering that the most popular question among Rotolactor tourists was "Which one is Elsie?" Borden's selected a good-natured cow to appear at the World's Fair as the first of a succession of Elsies. She was buried on the Walker-Gordon farm after her death in 1941, and her grave was yet another must-see component of a Rotolactor visit.

The doors of the Rotolactor remained open until June, 1971.

* * *

The second round attraction of my Jersey Shore childhood was the Neptune Music Circus. The original summer musical theater-in-the-round, under a circus-style big top, had been built in Lambertville, New Jersey in 1949 by the colorful actor and impresario St. John ("Sinjun") Terrell. The format proved hugely popular, and spinoff "tune tents," including the one in Neptune, quickly sprang up all over the country.

The closing of the Asbury Park Air Terminal three miles west of Asbury Park on Rte 66 provided Terrell with a perfect opportunity to bring this unique theater experience to the Jersey Shore. Inspired by Greek amphitheaters, steeply rising rows of folding chairs ringed the circular stage. The sets were of necessity simple and low, and stagehands could be seen running around moving scenery and props. A full orchestra ensured that nothing of the musical impact was sacrificed. Many then-unknown singers and actors got their start in this and similar venues.

The music circus craze turned out to be a bit of a flash in the pan, perhaps because of the inherent difficulties of staging such events, or perhaps because the world was ready to turn its attention to the rock music of the turbulent 1960s. By the 1970s most had folded their tents, though a couple forged on till early in the present century.

The Neptune Music Circus lasted only from 1952 to 1959, but its short lifespan came at just the right time for me—namely, my early adolescence, the mid-1950s, my last couple of years in Neptune. The dates are important because they marked the start of my parents' treating me something like a grownup. I remember the three of us getting all dressed up for an evening out and then gingerly picking our way across the muddy grounds in our good shoes, pretending we didn't notice. *South Pacific, Finian's Rainbow, The King and I, The Student Prince* (a major tearjerker for me), *Oklahoma, Brigadoon*—I remember all these performances with nothing but pure joy. My parents later bought the show albums—vinyl, as we now call them—and I danced around the living room belting out my favorites—"Bali Ha'i", "You've got to be taught", "How are things in Glocca Mora?", "Drink, drink, drink to Heidelberg", "OHHH-klahoma!"

A HOLIDAY GIFT FROM SEARS

sewing machine attachments; most of this stuff is now done electronically

hristmas was approaching in 2017 and excitement reigned among nostalgia buffs: The iconic Sears *Wish Book* was back! Well, sort of. It was mainly accessible online, but the retailers (absorbed by Kmart in 2005) assured us that Sears's best customers would also get a limited edition copy in the mail. Mine never showed up—guess it would take more than a couple of stops at the Sears watch repair counter to qualify for "best customer" status. That one year turned out to be its last gasp; the *Wish Book* is now gone forever. A year later, Sears it-self disappeared from our local mall.

* * *

Richard Warren Sears sent out the first flyer for his mail-order watch and jewelry business in 1888. He soon branched out into other merchandise, and by 1894 his modest mailer had evolved into a catalog. In 1896 the catalog grew even larger and started coming out twice a year. In 1897 a color section was added. An American tradition was born.

The Sears catalog proceeded to do for mail-order shopping what amazon.com later did for online shopping. In our house, as in many others, the arrival of the *Big Book* was truly an event. Twice a year, an enormous tome, 2-3" thick, was stuffed into our mailbox, in plenty of time for us to order our back-to-school wardrobe, our Easter finery, and our summertime shorts and halters. My sisters and I pored over its offerings, hoping they would make us look like the models in the pictures. Our choices bookmarked, we measured our chests, waists, and hips (which in fact were all about the same circumference in those days) and traced the outlines of our feet, all in a quest for the perfect fit. Our mom meticulously filled out the order form, which often spilled over onto two or three pages, carried it out to our mail-box, and raised the red flag to alert the carrier to the presence of outgoing mail. Then, biting our nails, we awaited the arrival of our precious cargo.

My all-time favorite was a dress that had also been purchased by the beauteous Ruth Ann, a classmate with an enviable put-together look—not a strand of hair would dare try to escape from her hip-length flaxen braids. The dress featured a quilted blue vest layered over a puffy-sleeved blouse adorned with red polkadots; an enormous red bow at the neck completed the look. It was actually singularly unflattering and made me look like Howdy Doody's twin sister (nothing was ever unflattering to

Ruth Ann); but I was proud to share my sartorial taste with this ethereal creature and always experienced a *frisson* of pleasure whenever the two of us showed up wearing "our" dress on the same day.

Although the clothing and toy sections were Sears's little-girl magnets, the catalog also featured sewing machines, sporting goods, musical instruments, saddles, firearms, buggies, bicycles, baby carriages, and oh, so much more. My parents swore by Sears products. As a young adult just starting out on my own, I inherited an ancient Kenmore vacuum cleaner from my parents when they upgraded to a newer model Kenmore. Mine looked like a torpedo, and old as it was, amazingly you could still buy any replacement part you needed. My husband and I used it till we decided we could afford something with more features—disposable bags and a shape better suited to navigating stairs were the big selling points, as I recall. It may have been a mistake; we've never had a more reliable vacuum cleaner.

The vacuum cleaner is long gone, but I still have my mother's Franklin Deluxe Rotary Model sewing machine in its original red-stained wooden cabinet, along with a set of bobbins and a box of lethal-looking attachments with names like the Multiple Slot Binder, the Five Stitch Ruffler, the Underbraider, the Edge-stitcher, and the Gathering Foot. Alone in an even bigger box is the Adjusting "Famous" Buttonhole Worker. The machine itself is an early electric model with a thigh-operated lever instead of a treadle. I learned to sew on it. Again, I stopped using it not because it was broken but because I wanted some fancier and more user-friendly features. I haven't plugged it in for half-a-century —I now use it to hold my inbox—but for all I know it still works.

They just don't make 'em like that anymore.

❊ ❊ ❊

Sometime during the fall, the Christmas catalog would arrive —the unofficial start of the holiday season, the stuff dreams were made of. This special edition concentrated in one place all the holiday-related merchandise from the *Big Book*—wax candles for trees, cards, ornaments, stockings, and artificial trees—along with additional items deemed to have an extra measure of gift-appeal. The first edition, in 1933, included the "Miss Pigtails" doll, a battery-powered toy automobile, a Mickey Mouse watch, fruitcakes, Lionel electric trains, a five-pound box of chocolates, and real live singing canaries. (Considering that children could legally be sent through through the mail after the introduction of Parcel Post in 1913 until someone got around to plugging that loophole, I guess we shouldn't be too shocked about mailing canaries.) Yes, it included many toys, of course, but even more pages were devoted to gifts for adults. In 1968, in deference to a sobriquet already in wide public use, the book was rechristened the *Wish Book*.

The Sears folks were well aware that their catalogs were a potential treasure trove for future anthropologists. In 1943, the year I was born, the *Sears News Graphic* referred to their catalog as "a mirror of our times, recording for future historians today's desires, habits, customs, and mode of living." Producers of Broadway shows and Hollywood movies still turn to the old Sears catalogs for help with retro styles and furnishings.

The *Big Book* was discontinued in 1993, but Sears was slow (too slow) to introduce online shopping and its fortunes dipped —Wishbook.com was launched in 1998, but the beloved print version wasn't retired until 2011, making its brief comeback in 2017.

Sears's mail-order business long preceded the Sears chain of retail stores, which didn't start opening until 1925 (oddly presaging Amazon's current expansion to brick-and-mortar retailing; been in Whole Foods lately?). There was a Sears store in Asbury Park, memorable to me because my father once "lost" me while he was shopping in (where else?) the hardware section. I wasn't really *very* lost, of course, but I was frightened enough that a kindly clerk took it upon himself to reunite me with my dad a few aisles away. (My mother never, ever lost track of her children in a store!)

I can't remember why we didn't shop in the store more. I mean, why did we order our clothes from the catalog if we could go and try them on in the store? Perhaps there was no children's clothing department? Or only a limited selection of styles and sizes? Or no clothing at all? All I know is that it never occurred to us **not** to catalog-shop. Perhaps it's the 20th century equivalent of ordering something on amazon.com that I could easily pick up in the drugstore.

* * *

In 1908 Sears outdid itself and started selling kit homes designed to be put together on-site by the owner—a process that clearly took "assembly required" to a whole new level. About the only precondition was that the buyer live near enough to a railroad line so that the lumber, asphalt shingles, plaster and lath (later drywall), windows, flooring, hardware, and everything else needed for the project—around 30,000 pieces in all, including 750 pounds of nails and 27 gallons of paint, enough for 3 coats on the exterior—could be delivered in boxcars. Each part was numbered and had to be matched up with the extensive instructions and blueprints that came with the kit. Sears even provided financing; heating, plumbing, and electrical systems were extra. Decorating advice was available on request.

Sears was not the only purveyor of kit houses nor even the first, but it was the largest and most diversified. Several editions of the 1908 *Modern Homes* catalog were issued, ending up with a selection of more than 40 house designs, with prices ranging from a few hundred to a few thousand dollars. By the time *Modern Homes* came to an end there were 447 models in all, according to the Sears Archives, in different sizes and architectural styles, stratified into three lines to accommodate different budgets. Some were only offered once; other more popular models were evergreens that reappeared year after year.

After absorbing many foreclosures during the Depression, Sears discontinued financing in 1934. In 1940 the last of the *Modern Homes* catalogs was published, but kits continued to be sold through 1942, including designs from the 1940 issue as well as new designs that had never appeared in the catalog. By then an estimated 100,000 had been built, many of which are now listed on the National Register of Historic Places.

Before a recent visit to Shark River Hills, I emailed my childhood friend Bill to ask if he had any "then" photos to pair with the "now" photos I was planning to take. He responded with two photos of his family's home on South Riverside Drive, taken in 1949. His father later dug out the crawl space beneath the porch to extend the basement and enclosed the porch itself to provide more living space, so the house looks quite different today. But those changes were made after the Stodolas left Shark River Hills.

In his narrative about the house's history, Bill also mentioned —much to my surprise—that it was one of several summer homes built in the 1930s by a developer from Sears kits shipped by rail to Belmar and then transported to Shark River Hills for assembly. Unlike his family's comfortable two-story home, said Bill, most of these summer houses were smaller, single-story affairs.

"That's the story I heard anyway," he added.

Unfortunately, it is not always easy to identify Sears homes. A few years after Sears stopped selling kits, all sales records were jettisoned. Some homeowners did not want it known that they lived in kit homes and destroyed the evidence. Sears offered reverse floor plans and encouraged buyers to customize—to swap out the siding material, to add a dormer—increasing design variability. Competing kit home manufacturers often copied design elements from one another, which can also contribute to confusion.

Still, there are a few markers—stamps on the lumber used for framing on houses built after 1916, shipping labels, a small circled "SR" cast into the lower corner of the bathtub in houses built during the 1930s, etc.—that can help to support or rule out a house's pedigree. Paperwork found in the home and legal documents can also provide clues, as can comparison with published house plans. And of course, unless the house was built between 1908 and 1942, it cannot be a Sears home. (See Rosemary Thornton's fascinating books on Sears houses for more help in verifying authenticity.)

Although New Jersey boasts many kit houses, I have yet to uncover any more information about the would-be Sears homes of Shark River Hills. It is well to bear in mind that many family legends are simply that, legends, and more people *think* they own Sears houses than actually do. Perhaps an architectural historian will one day step forward to corroborate or definitively disconfirm Bill's claim, and possibly even point out some additional Sears homes that might, just might, be hidden in plain sight in Shark River Hills.

B. KID STUFF

SALLY

in the tire swing

Sally and I met in around 1944, which probably makes her my first real friend. I can't remember a time when I didn't know her.

When you're very young, everyone seems to be one-of-a-kind and larger than life, and my memories of Sally are almost psychedelically vivid. They are largely confined, however, to our first thirteen years, before my family moved away from Neptune. I was just as solipsistic as the next kid, so most of my childhood recollections of Sally are actually at least as much about me as they are about her. There was the time I told Sally the "facts of life" after having The Talk with my mother, for example, and the time I gave Sally the lowdown on Santa Claus—both of which landed me and my big mouth in deep hot water with Sally's mom Mary Jane!

When the Everses first moved from Long Branch to Shark River Hills, they lived so far away that you had to drive to get together. But when I was around five we moved to Pinewood Drive, and not too long after that the Everses moved to Riverside Drive near its intersection with Pinewood Drive, just a couple of blocks away. From that time on, the children from the two families were inseparable. Mary Jane became my "other mother," and Sally later told me how warmly she remembered my mom. (I was always a tiny bit afraid of Jim, Sally's dad!) The black cat with whom my own children grew up was named after the Everses' black cat, Rasputin.

Life for two little middle-class White girls living in Shark River Hills in the 1940s was in many ways quite idyllic. We went out to play in the morning and often didn't show up again till it was time to eat. (If moms permitted that today they'd probably have Social Services on their doorsteps, but somehow we managed to thrive.) Here are some of the things Sally and I did together: rode our bikes down the Big Hill (now called Snake Hill; how steep it seemed then! and we did it "no hands") and on to the Cracker Barrel to buy a candy bar or an ice cream cone; took scary safaris in the deep, overgrown ravine next to the Everses' house; hiked in the woods in search of acorns (plentiful) and lady slippers (rare); painted our initials on the backs of box turtles in hopes they'd return the following year; wiled away the long lazy summer days on our front porch playing card games and board games. And of course no account of a Shark River Hills childhood in the 1940s and 1950s would be complete without a mention of the Clubhouse, the community center near the river where everyone young and old gathered for activities ranging from knot tying classes to potlucks to beauty contests—not to mention the Tom Thumb Wedding at which a boy named Eddie Jackson and I impersonated Sally's parents.

Good-sized families were the rule rather than the exception in those days. At first there were only Sally and me and the "Little Kids," Sally's sister Barbara and my sister Leslie, whom we played with, fought with, hid from, and (I fear) teased mercilessly. (Sorry, guys!) The Evers clan eventually expanded to include Sally Jane, Nancy Jane who died in infancy, Barbara Jane who played with my sister Leslie, Billy (who understandably believed his name to be Billy Jane), and Helen Jane who played with my sister Sherry. Five years after Sherry, my brother Bob arrived, and sometime around then along came Susan Jane, with whose birth Sally finally outstripped me one and for all in the sibling department—four Stodolas, five Everses!

Sally was an adorable little girl, with straight strawberry blond hair, maybe a little tomboyish—a real kid's kid. When we were little, she was always quite a bit taller and heavier than I was, and I sometimes ended up with her hand-me-downs. When I left Neptune I was still a lot smaller than she was—so it was a surprise to return later for a visit and discover that I towered over her by three or four inches. She had stopped growing but I had not!

Later in childhood we both branched out and forged other more mature and more complex friendships. But for as long as the Stodolas remained in Neptune, the strong bonds persisting from the time when we were almost sisters and essentially shared two mothers never entirely disappeared. Although then, as now, interactions among groups of girls could become fraught and even stormy, I can't remember ever seriously quarreling with Sally. We must have had our share of tiffs, but unless someone or something else competed for our attention, we just sort of naturally gravitated to each other.

In 1956, the summer after we graduated from 8th grade at Summerfield Elementary School, my family moved away from Shark River Hills, and at that point Sally and I pretty much lost touch with each other.

Then, as I started to plan a trip to Long Island in September of 2010 to attend my 50th high school reunion, I realized that Neptune High School's reunion was just a week earlier and only 90 miles away from Long island. Sally warmly welcomed my bright idea of attending as an "honorary member" of the Neptune High School Class of 1960 and reconnecting with old acquaintances from Summerfield. Now that we were back in touch, we started swapping Mary Jane and Elsa stories and reminiscing about days gone by. When Mary Jane died not long after, Sally emailed me to say: "Mother passed away on Sunday. It was peaceful and I'm very grateful for that. Just wanted to let you know since she was a part of your life, too."

So I was shocked speechless when I arrived at the reunion banquet, asked for Sally, and was told she had died. My immediate reaction was that there must be some mistake—they must be talking about Mary Jane's recent death. But no, sadly, it was in fact Sally, who had succumbed unexpectedly to a sudden, brief illness.

A sobering reminder to cherish every interaction as though it were your last, because it just might be.

A VISIT TO THE EASTER BUNNY

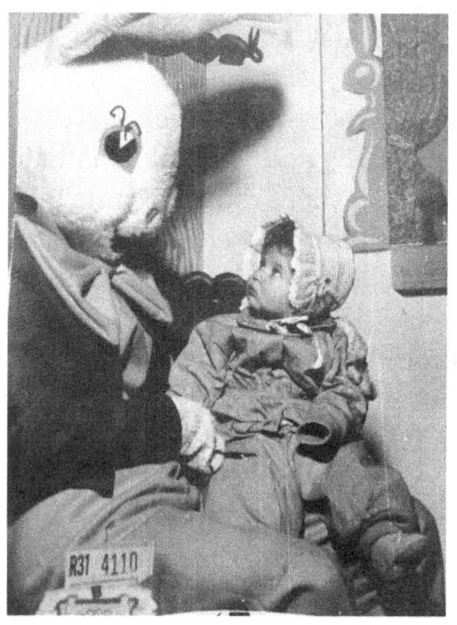

S herry, the younger of my two younger sisters, sent a "Happy Easter" message to our family a couple of years ago, attaching a photo of herself, age eight months, sitting on the Easter Bunny's lap. The look on her face is a perfect combination of bewilderment and fear, and her little left hand is completely drawn up into her sleeve. "The Easter Bunny may have eaten my hand; no wonder I look somewhat skeptical," she captioned the photo. "Cindy and Leslie also have pics from this visit to the hutch," she added.

She's right, I still have my own photo of that outing, also taken in the Spring of 1949, shortly after I turned six. A few years ago I posted it on Facebook on Easter Day, and one of my friends observed that the Easter Bunny looked a little creepy. I can no longer look at the photo without thinking of that comment, and Sherry's "missing hand" photo only serves to increase the vague sense of uneasiness the images evoke. My brother, who missed the Easter Bunny photoshoot for the excellent reason that he hadn't yet been born, opined that the registration number (visible on both photos) just adds to the creep-factor. (Fortunately for our tender little psyches, I don't think any of us ever believed in the Easter Bunny with quite the same uncritical fervor as we did in Santa Claus.)

Bunnyland, as I recall, was located in nearby Asbury Park in Steinbach's, which at that time billed itself as "the world's largest resort department store." The Asbury Park Steinbach's, founded in the late 19th century, was the flagship store of what eventually became a chain, with branches dotting the Jersey coast. The store later fell on hard times after racial tensions flared in Asbury Park and was permanently shuttered in 1979, but when I lived there Steinbach's was in its heyday. The trapezoidal "flat-iron" style building that I recall, on Cookman Ave—a real eye-grabber—had been built in the 1930s with four floors and a basement; by the time I arrived on the scene, a fifth floor and a clock tower had been added.

I actually found Steinbach's a little intimidating, with its multiple floors of merchandise and elevators run by uniformed attendants, and I don't recall ever going there alone. I was more comfortable in the "five-and-dime" variety stores like Woolworth (sort of like what are now, what with inflation, called "dollar stores"), with at most two floors, connected by moving staircases that offered the illicit thrill of reaching the second floor by running very fast up the down escalator. But when my mother wanted to go a little more upscale—to buy her lingerie, for example—she went to Steinbach's. And since Steinbach's had the only Easter Bunny in town, that's where she took her three young daughters to pose with the big guy with the floppy ears (the cute photos—available for purchase, of course—being the main object of the encounter, since wish lists and naughty-or-nice issues weren't part of the Easter narrative).

❊ ❊ ❊

Easter was a happy holiday for us, a day that signaled the arrival of Spring—a time of rebirth, a time when we could don our fanciest finery without freezing our bare legs, a time when we could start dreaming about summer vacation, about hunting for lady slippers and catching box turtles.

It also brought out our latent artistic impulses. When it came to the dyeing and decorating of eggs, families aligned themselves in two camps, those who hard-boiled their eggs and a smaller faction that hollowed them out. We were in the latter camp (and for that reason I still have eggs decorated by my daughters when they were children). We punctured both ends of the eggs and then blew on one end, somehow avoiding the twin hazards of salmonella from placing our lips on the unwashed eggshell and apoplexy from the eye-popping effort required to blow the entire contents of an egg through that minuscule opening. (Trust me, this is no easy task, especially if you want the holes to remain small and inconspicuous. I usually ended up cheating and making the pinholes a little larger. It also helps to break up the yolk with your neeedle.)

The night before Easter, my mother filled our baskets (each of us had her own, recycled and restocked year after year) with Easter grass, the eggs we'd decorated, and a mouthwatering assortment of chocolate bunnies, jelly beans, and gumdrops. (No peeps—those weren't mass produced till 1953.) My own favorites were those large crystallized sugar eggs with windows looking into a miniature alternative universe—not because of the sugar (though that was delicious when the confection finally dried out and crumbled) but because I liked to fantasize about crossing through that window into my own little Wonderland. You can still buy these eggs but somehow the interior landscape doesn't seem nearly as elaborate or compelling. Or maybe it's I who have changed.

After the baskets were assembled, my mother hid them, and the next morning we had to search for them. And by "hid them" I mean she *really* hid them, and never in easy, obvious places. One year she hid Sherry's basket in a seldom-used closet and swarms of ants found their way to the candy, so Leslie and I had to share our goodies with our little sister. Life's like that sometimes.

Later in the day, having already stuffed ourselves with candy, we gorged on ham with pineapple slices studded with cloves, or perhaps on roast lamb with green mint jelly—both typical Easter fare of the era. Only Thanksgiving was a more tradition-laden dinner. For our friends who gave up something they cherished for Lent, the Easter feast ended forty days of deprivation. We enjoyed the goodies without the deprivation. Life's like that sometimes, too.

A JERSEY SHORE CHRISTMAS

Christmas 1945, just before Project Diana

As soon as Thanksgiving was over, the Christmas season began in earnest. My mother made two fruitcakes, one dark and one light—the source of my own compulsion to bake fruitcake every year whether anyone likes it or not—which she doused liberally with brandy once a week until Christmas.

Around the same time, my father scheduled the family Christmas card photo shoot. The cards began in 1943, when I was 10 months old. There was no card for 1944, but after that, the series remained unbroken; some years there were actually two different versions. Our last Christmas in Shark River Hills was 1955; the cards continued for a few more years at our new home on Long Island. My father clearly loved this annual ritual and had endless patience in his quest for perfection. He set up a tripod and lots of spotlights, and some years he also draped the room with sheets to provide a neutral backdrop. Elaborate scenarios were developed (clutching our pets; reading to the younger children; sitting at an easel pretending to paint or at the piano pretending to play). His children didn't always share his enthusiasm (as unfortunately can be seen from our glum expressions in some of the photos—re-do's weren't as easy as with an iPhone!). We had to sit very still, with our pasted-on smiles, and the spotlights made the room hot as an oven. Today, of course, I am inordinately grateful not only for the memories my father made for all of us by staging this event but also for the photographic record of how we all grew and changed and added to our number over the years. Wish I could tell him so now.

A couple of weeks before The Day, we decorated the house. My mother had a creche that we set up on the mantle, and even though we weren't particularly religious, I loved the baby Jesus and the whole family tableau. (I still do.) Right below them hung the stockings awaiting Santa's attention—an interesting juxtaposition of Christian and pagan symbolism, though not one we thought much about at the time. We also had a little cardboard village with colored cellophane windows and holes for Christmas lights, which she arranged on the piano. A wreath went up on the door.

My father set up the tree in a semi-finished "game room" in the basement, near the pingpong table (and later a pinball machine that he bought third-hand from someone, or perhaps from a pawnshop—all the Stodola girls inherited his love of thrifting!). The beloved box of ornaments came down from the attic, and we competed to be allowed to hang our favorites. Another predictable squabble was sparked by the silver foil icicles: I liked to hang them slowly and painstakingly so they would look like real icicles, while my sister preferred taking clumps of the stuff and flinging them at the tree.

These mini-crises resolved, we artfully arranged our gifts under the tree—all but the ones from Santa, who didn't visit till Christmas Eve after all of us (including, we supposed, our parents) were sound asleep. (I'll never forget how proud I was when I was deemed old enough to be dropped off at Woolworth in Asbury Park to do my own Christmas shopping, using the money I saved from my allowance by depositing fifty cents per week in a "Christmas Club.")

* * *

No portrayal of Christmas in Shark River Hills, or any other holiday for that matter, would be complete without mentioning the firehouse on Brighton Avenue. The firehouse was more than just headquarters for the volunteer fire department, it was the beating heart of the community, serving as a meeting place for scout groups and other organizations and an event center for community parties and celebrations. My mother was a member of the Ladies' Auxiliary, even though my father was not a firefighter. Shortly before Christmas, the local children donned their holiday finery and gathered for a party during which I can't remember what happened. Candy? Santa Claus? Definitely a group photograph. For a while the photo from 1953 was floating around Facebook, and identifying the 73 in attendance became an obsession among the SRH crowd, including a circulating excel file that ended up with about 3/4 of the names filled in. Both my sisters were there, my BFF Joyce, and practically everyone else I knew (plus a few I didn't). Where was I? I guess I must have been ill that evening; surely I didn't have anywhere else to go!

* * *

My parents always hosted an Open House for all the neighbors (children and adults alike) on Christmas Eve. During the preceding week, we made dozens and dozens of cookies—including sugar cookies, which we decorated with red and green sugar, and Tollhouse cookies, made by following the recipe on the back of the Nestle's package, which magically and consistently produced the Best. Ever. Chocolate. Chip. Cookies. Out came the two fruitcakes, dark and light, for one last splash of brandy. Just before the party my mother prepared two batches of eggnog, nonalcoholic for the kids and most definitely alcoholic for the adults. We donned our Christmas finery and were allowed to stay up late. Years later, one of the neighborhood girls told me she was so inspired by these parties that as an adult she has always given a Christmas Eve Open House of her own.

After all the guests had gone home, we put on our foot pajamas and snuggled up to listen to "The night before Christmas"; then it was off to bed with us so we wouldn't be too tired the next morning. Not a problem for me; I was almost always the first one up. But we still had to wait for our parents to get up before we were allowed to go down to the basement and start opening our gifts—which seemed like forever but was probably more like half an hour. We had made our Christmas lists and we usually got exactly what we'd requested, plus lots of other stuff. One year I asked for a doll I coveted but stipulated that I wanted her wrapped so I could be surprised when I opened the package. This turned out to be a bad call, since she ended up with a bad case of "wrapping paper head" that never quite went away no matter what I did.

When I was around five, my father bought us an electric train. After we went to bed on Christmas Eve, he stayed up long into the night laying the track so that the train would disappear down a hallway and a couple of minutes later reappear through the dining room. That year he, not I, was the first one out of bed on Christmas day. The train was a big hit with all of us but no one was more excited than my dad. If you want to make an engineer happy, just give him a model train and a whole day with nothing else to do but play with it.

QUEST OF THE MAGI

T his is one of my favorite photographs, of one of my happiest childhood memories—the annual reenactment of "We Three Kings" by my sisters and me, accompanied on the piano by our grandfather. It took place sometime between Thanksgiving and Christmas, at my grandparents' home in Oakland, New Jersey, some 70 miles from Shark River Hills.

In the back of a large closet, my grandmother kept a trunk crammed with the makings of just about any costume a child could desire—exotic scarves and shawls, colorful lengths of tulle and velvet, and opulent fabrics threaded with silver and gold. I never saw my grandmother sew, so I suppose these treasures were gleaned from the wardrobes of her various thespian activities. Her grandchildren were always welcome to delve into its depths, strewing its contents around the room as we searched for just the right pompom or scrap of lace to complete the look we were striving for.

At no time was our fervor for such revelry greater than when the holiday season rolled around.

Because their three-ness matched our three-ness, my two sisters and I felt a particular affinity for the Three Wise Men. When *Amahl and the Night Visitors* debuted on Christmas Eve of 1951, we quickly adapted our own hilarious (to us) and rather raucous version, prancing around the house and belting out "Mother, Mother, Mother come with me!" at the top of our lungs.

Not even Gian Carlo Menotti's charming opera, however, could compete with our perennial favorite, the splendidly dramatic old carol "We Three Kings." Each of us had her own favorite king and her own favorite verse to sing. Mine was Balthazar; for some reason I found his lugubrious description of his gift perfectly irresistible: "Myrrh is mine; its bitter perfume / Breathes a life of gathering gloom / Sorrowing, sighing, bleeding, dying / Sealed in a stone-cold tomb." I hadn't the faintest idea what myrrh was, of course, but then I didn't know what frankincense was, either. No matter, we all knew they had to be something pretty special since they obviously had to equal the value of the third gift, gold.

Once we had perfected our costumes, committed our verses to memory once again, and duly rehearsed, we herded our captive elders into the living room, and my grandfather, after considerable coaxing, sat down at his Steinway. He had once been a concert pianist, but by now his skills had sadly atrophied from disuse and arthritis. I now have an inkling of how painful this must have been for him, but at the time I was completely oblivious. I only hope he knew how grateful we were for his critical contribution to our production.

And this, *this* is the magic moment captured in the photo—the performance just about to begin, Sherry and I trying our best to look serious as befitted the solemn occasion, the irrepressible Leslie grinning from ear to ear, our long-suffering grandfather seated at the keyboard, chomping on his ever-present cigar. My father was undoubtedly the photographer. Truly a Stodola classic.

THE TOM THUMB WEDDING

A while ago Bill Evers sent me a photo of the wedding party of a "Tom Thumb Wedding" in which I partici- pated at the Shark River Hills Clubhouse—thus doub- ling my collection of photos from this event. (I've since acquired a third.) Bill remarked at the time that the concept of a mock wedding for children was foreign to him. I too think it odd—not only from my adult perspective, but I actually remember finding it difficult to wrap my brain around at the time. (Remind me, why are we doing this?)

So I did what I always do at such moments—Googled—and found that the event was part of a tradition inspired by the real-life wedding of Charles Stratton, a dwarf who performed and traveled with PT Barnum as "General Tom Thumb", and Lavinia Warren, only a few inches taller than her groom. It is almost impossible to recapture the excitement generated by this event, which was probably the closest American equivalent to a royal wedding. Details about the bridal gown and trousseau were bruited about in the press, costly gifts were forwarded from around the world, and socialites vied (and paid) for invitations to the ceremony at Grace Episcopal Church in Manhattan on February 10, 1863. The wedding and its associated festivities provided a war-weary nation with a welcome respite. President and Mrs. Lincoln themselves became involved, hosting a lavish reception in the diminutive couple's honor and inviting them to honeymoon in the White House.

Shortly thereafter, reenactments of the wedding started being staged by schools and churches all over the country, as youth activities and fundraisers, and also to teach children "values." The practice seems to die out from time to time, only to spring back into vogue. Even today, reports and photos of recent Tom Thumb weddings can be found in the press and online. (A modern twist: A GoFundMe site appealing for contributions to support a boy's candidacy for the role of groom!) Small children (usually under the age of ten) formed the cast, although sometimes the minister was played by an adult. In the one in which I participated, which took place on July 27, 1950, all the characters were played by children, minister included, with music provided by a couple of slightly older kids. Participation was maximized by including not only a full wedding party but also guests—prominent members of the community also played by children.

I won't say it wasn't fun. My sister Leslie played a bridesmaid and I played Bill's mom, Mary Jane. I think she supplied the silky navy blue dress I wore, heavily altered to fit a seven-year-old; possibly the hat was hers as well. I felt very dignified, though in retrospect I wonder what Mary Jane made of her pint-sized doppelgänger, bedecked in her finery and tottering around in high heeled shoes that were several sizes too large. My "husband," Jim

Evers, was played by a boy named Eddie Jackson, whom I didn't know very well and might not even remember except for that chance coupling (in which, by the way, no consent or choice was offered—all the decisions including casting were strictly top-down).

* * *

Did this spectacle provide anything beyond amusement for the participants and spectators? Did it instill an appreciation of commitment and responsibility? Did it help to prepare us for the adult world? Did the pomp and pageantry promote community spirit? Did it, on the other hand, encourage us to cling to a stereotyped set of middle-class mores and discourage openness to other life choices? Did these elaborate charades have the effect of reducing what is meant to be a solemn occasion to a circus sideshow?

According to historical novelist Melanie Benjamin, blogging in the *Huffington Post*, Lavinia Warren Stratton "never really knew how to view these staged weddings; were they tributes to her great love? Or mockeries?" If Lavinia herself was puzzled, I guess I can be excused for my own mixed feelings about the Tom Thumb wedding.

OUR PETS

We always had pets—at least one dog and one cat, and often more. A few of them show up in the famous Stodola Christmas cards.

I am ashamed to confess to having possibly contributed to the Stodola cat census via an acquisition method that could best be described as "Look what followed me home." Never mind that I was covered with scratches from the process of persuading a reluctant feline to "follow" me home. I suppose I may have ended up keeping one or two of these kidnapped unfortunates; most, however, were truly feral animals that were mercifully (for all concerned) released back into the woods.

* * *

Before I was born, my parents adopted a kitten and pup duo my mom called Nip and Tuck. I don't really remember them, but I do know that was the last time my mother ever had a chance to name our pets. After that my father took over, and he reigned supreme till we kids were old enough to join in the name game.

None of those cute pair names like Nip and Tuck or Salt and Pepper or Whisky and Soda for him! Rather, finding something catchy to call our pets was only a starting point for the prolonged discussion by which, through some cabalistic process, we eventually arrived at their true titles. So it turned out that Perky the tabby was really Percopolis, Penny the little car-chasing spaniel was Lady Penelope Penny van Pennysworth, Laurie the goofy boxer was Laurelita von Sniffnwoof, and Goldie, our gentle giant of a yellow tom, was actually John Timothy McGoldrick.

My dad came by this practice honestly. I can't remember a time when his own mother didn't have a calico cat capable of prolific reproduction, and although we always knew each in this long succession of creatures as "Orrie," their collective official name was Aurora Borealis. My grandmother charged $5 for each of Orrie's kittens, claiming they were much easier to place if you sold them than if you gave them away; somehow, to my mother's eternal amazement, my grandmother succeeded in selling out every litter.

My mother always claimed not to like pets very much but anyone who watched her with them knew better. (Not that we children could believe anyone could possibly dislike the little guys. It wasn't till I was an adult with kids of my own that I learned how much work they could be, especially dogs, even if you adored them. Guess who sheltered me from that knowledge!)

One dog-related task that I always participated in enthusiastically, however, was a regular Saturday morning session that somehow turned into a bizarre father-daughter bonding ritual.

No one had ever heard of Lyme disease then—perhaps it wasn't even a recognized diagnosis yet—but living at the Jersey Shore meant lots and lots of ticks, and the ones we knew best were dog ticks. When the season started in the Spring, my father filled a glass jelly jar with kerosene and then sat on the cellar stairs with a dog between his knees, tweezers in hand and his daughter at his side, watching with abject fascination. As the season wore on, the jar would become filled, with both the athletic little black ones that had barely had a chance to latch on and the large bloated white ones that had been gorging on dog blood for days. By the end of the season my father would happily display his trophy collection to anyone unwary enough to feign an interest in the process.

Of course we had a succession of nameless fish and turtles, but can you call a box turtle a pet? Every summer we found a few good-sized box turtles lumbering along the roadside—easily distinguished from their surly snapping cousins by their much sweeter faces and dispositions. We kept each one in a box for a few days, painting our initials on their backs and feeding them lettuce, then released them and waited to see if they would return the following year. That seldom happened—though I once found one with someone else's initials!

Other non-pets included the fireflies we kept in jars by our bedside, watching dreamily as they lit up the room after dark.

Beyond that, my many rather elaborate efforts to catch and adopt wildlife were uniformly unsuccessful. For some reason a squirrel was at the top of my wish list. No one ever explained to me that our pets were bred to be infantilized, and that wild animals could not readily be tamed; or if anyone did I didn't listen.

※ ※ ※

One of my all-time favorite pets was Archie the parakeet. Parakeets (or budgerigars) are small parrots living wild in the drier regions of Australia. They have been bred in captivity since the 1850s, but for some reason a parakeet craze swept the nation in the 1950s and early 1960s. Our Asbury Park Woolworth was literally atwitter with colorful creatures awaiting adoption —green (the wild type, like Archie), but also blue and yellow and white that had been bred for variety. I longed for a parakeet the moment I saw their comical little faces, and since in our house longing was usually the prelude to receiving, sure enough, the coveted cage awaited me on Christmas morning.

Although we were familiar enough with marine creatures, birds had never before been part of our family menagerie, so without warning I was introduced to the mysteries of cuttlebone, sandpaper tubes to cover perches, how to hand-feed a parakeet without getting pecked, and (less alluring) the weekly cage cleaning requirement.

But what really excited me was that you could supposedly teach your parakeet to talk. Forget about canaries; I wanted my bird to converse with me, not serenade me. Unfortunately, either Archie wasn't the brightest budgie in the flock, or more likely we weren't systematic enough to train him properly. For whatever reason, we never succeeded in turning Archie into much of a raconteur.

He did, however, have one notable verbal accomplishment. In those days when you wanted to make a telephone call, you picked up the receiver and the operator (a real human, and always a woman) said "Number, please." Almost all our numbers began with "Asbury Park two," the local exchange, followed by four additional digits that uniquely specified the recipient. Archie's cage hung right over the telephone, and Archie became quite adept at saying the three little words he heard most often, "Asbury Park two."

Sadly, Archie came to a bad end at the claws of one of our mama cats. I was devastated and insisted on wearing a black armband to school the next morning.

* * *

The invention of kitty litter in 1947 by a man named Edward Lowe changed cat ownership forever, in ways that went far beyond which side of the door the cat spent the night on. Before kitty litter, cats were workers who earned their keep by keeping rodents at bay; since kitty litter they have become fur babies who, if we're lucky, curl up with us at night.

In my childhood home and I'm sure in many other homes of the late 1940s and early 1950s, however, this new invention hadn't quite caught on yet. I also don't remember the practice of routinely neutering male cats, though females were generally spayed. Consequently, there was no such thing then as what we would now call an indoor cat, and life with a tom involved a lot of patching up.

So it was with John Timothy McGoldrick, son of one of the many Orries and a wondrous beast with totally contrasting outdoor and indoor personalities. Outdoors, birds feared him, and rightly so. Indoors, he drooled like a baby and happily let Archie perch on his head, purring all the while.

He did, however, have a disconcerting habit of disappearing for days at a time, and as he grew older his absences grew longer. The day my parents finally left the house on Pinewood Drive to move to Long Island, Goldie was nowhere to be found. All the neighbors were asked to watch for Goldie and notify us immediately if he turned up, but he never did.

TONI'S DOUBLE LIFE

Toni looking jaunty in her jumpsuit, made by my mother

Of all my childhood toys, only two remain with me today: my Hazel-Atlas "modern tone" tea set (supplemented by eBay purchases to replace pieces that went missing somewhere along the way) and my beloved Toni doll.

My Toni is a model P90—that is, 14" tall (the smallest Toni; the dolls also came in several larger sizes.) She has blue sleep-eyes lined with a fringe of upper lashes and painted lower lashes, a slightly pouty red mouth, a jointed body, and a platinum wig glued to her skull. Sadly, one of her hard plastic pinkies has broken off, but she is otherwise in pretty good shape. She was manufactured by the Ideal Toy Company sometime between 1949 and 1953, so I couldn't have been more than ten when she and I began our long journey together, and more likely closer to six. I suppose I should display her on a stand and try to protect her from further damage, but until recently my granddaughter loved to play with her and I love watching my granddaughter at play, so Toni's fate has been to lead the rough-and-tumble life of a child's toy and not the pampered retirement of a collectible gathering dust on a shelf.

In addition to Toni herself, I have a whole wardrobe of clothes whipped up for her by my mother on her trusty old Franklin, using patterns from Butterick or McCall's and material left over from earlier sewing projects. In fact, for a long time I had two of each outfit, one for Toni and one for her red-haired sister Nancy Lee, a slightly larger Arranbee doll with which I foolishly parted when her arms and legs came off, not knowing how easy it would have been for any "doll doctor" to reattach them. I no longer have Toni's original outfit, detracting from any value she might retain as a collectible, but the ones my mother made for me are far more precious.

* * *

Although to the untutored eye she resembles many dolls of her era, my Toni doll has a dirty little secret: In addition to being a charming toy with lots of little-girl appeal, she was also a promotional gimmick for Toni Home Permanents. Designed by the renowned German immigrant doll sculptor Bernard Lipfert, she came packaged with her own home permanent kit that included a sugar-water permanent solution, end papers, curlers, and a comb. Yes, you could actually perm Toni's nylon wig—though if you did it too often her hair could turn into an irrecoverably sticky mess.

In the past women had styled their hair with curling irons and, well, just plain irons, which all too often left their homes reeking of scorched hair. So the development of the permanent wave in the early twentieth century was embraced by many women, especially after the less elaborate cold wave was invented in 1938. But even though the alarming Rube Goldberg machines and strong heat of the old-style perms were no longer needed, the cold wave process still involved serious chemical changes to the protein structure of the hair and required six to eight hours in a salon.

The home perm, pioneered by the Toni Home Permanent Company of Forest Lake, Minnesota, was thus a breakthrough product, offering a cheaper alternative to the salon perm and bringing hair styling back home again. It also made beauty a social occasion, with Toni parties becoming popular among both teenagers and adults.

In 1948, the Toni Company was acquired by Gillette, an early step towards diversification for Gillette and also the start of more aggressive marketing of Toni products. The best-known (and clearly fraudulent) ad campaign featured photographs of wavy-haired identical twins and famously asked, "Which Twin has the Toni?" The twins themselves sometimes appeared in touring shows that invited audience participation in making the "correct" identification.

Less attention has been paid by historians of advertising to the innovative process by which the Toni doll's creators sought to engage the consumer by appealing to her children. I was blissfully unaware not only of being targeted by the advertising industry but also of the delicious irony of having wheedled my mother into paying for this privilege.

Long hair and heavy machinery don't mix, and the employment of women in factories during World War II led to a call for shorter hairstyles. (The sultry actress Veronica Lake famously cut off her "peek-a-boo" locks to help promote workplace safety, and although her career suffered as a result, her sacrifice led to a measurable decline in industrial accidents.) The home perm was well adapted to these new styles and helped prolong a preference for soft waves or curls, often swept away from the face or paired with bangs.

Sleeker straighter hairstyles did not become fashionable till the 1960s, and my poor sister, who had the most adorable straight hair and bangs, was subjected in the name of beauty to a series of Tonettes, a Toni home perm intended especially for little girls. My own hair, on the other hand, was *too* curly. I desperately wanted to let it grow long, but my soft-hearted mother couldn't bear the anguished tears produced by combing out my sausage curls and repeatedly dragged me off to a beauty parlor for a cringe-worthy do called the "cap cut."

I'm not sure when creme rinse—a thin white liquid that magically softened and detangled thick curly hair—became readily available, but when it did, it changed my life. (It now seems to have morphed into much thicker "conditioners" that moisturize, volumize, and decrease frizz. Sorry, folks, it's not the same stuff.) Once I discovered creme rinse, I let my hair grow out and didn't trim it back to shoulder-length until I turned forty.

ME AND MY UKULELE

I received a ukulele for Christmas one year and learned to play it well enough to give me pleasure and avoid annoying those around me. Compared to the other instruments I took up as a kid (or more accurately had thrust upon me)—the piano and the saxophone (for which my father paid $25 in an Asbury Park pawn shop)—my ukulele was, well, quiet. And while playing it at a virtuoso level would have taken more practice than I was ever willing to throw at it, and probably more talent than I could muster, it wasn't that hard to tune it (G-C-E-A, with the memorable lyrics "my dog has fleas") and to learn enough chords to make my way through a few familiar songs. "On Top of Old Smoky," all umpty verses—even without Googling I can still dredge up five or six—was my hands-down fave.

But what was a little girl in a small coastal New Jersey town doing with an exotic Hawaiian stringed instrument? Luckily enough, I have a photograph of myself clutching my ukulele—slightly blurry, but clear enough to permit further research on the topic.

* * *

The ukulele is a plucked stringed instrument belonging to the lute/mandolin family. Although it is sometimes thought of as quintessentially Hawaiian, along with hula dancers in grass skirts and leis, it was actually developed in the 1880s by Portuguese immigrants who came to Hawaii from Madiera and Cape Verde, based on similar guitar-like Portuguese instruments. Its place in Hawaiian music and culture was cemented by the support of King Kalakaua, who enthusiastically promoted the instrument and made it a central part of performances at royal gatherings.

The U.S. mainland discovered the ukulele in 1915, at the Hawaiian Pavilion of the Panama-Pacific International Exposition in San Francisco. Being cheap, portable, and relatively easy to learn to play passably well, the uke quickly passed into the popular culture, appearing in film and becoming a mainstay of jazz, country music, and American song.

The life trajectories of two men converged to turn the ukulele into the musical phenom it was to become in the post World War II era:

The first was Arthur Godfrey, folksy host of a series of eponymous radio and then television variety shows (and also, incidentally, an avid ham—K4LIB). My Aunt Phyllis, my mother's younger sister, always claimed to dislike Arthur Godfrey because, she said, he laughed at his own jokes, but whenever his show came on the air, she poured herself a cup of coffee and sat down to listen. We often spent several days with Aunt Phyllis and her family in Worcester, Massachusetts during our summer vacations, and when "the Old Redhead" (as he called himself) came on, we stopped what we were doing and joined her in her kitchen. He was famous for poking good-natured fun at his sponsors' products. The off-script commercial I remember best was his test of the proposition that Lipton teabags could be re-used for several days in a row. The second day came and went without too much comment, but by the third day his reaction was simply "blecch." He got away with it because it was known he would not accept a sponsor whose products he didn't person-ally enjoy, and because (bottom line) it sold those products.

"Arthur Godfrey's Talent Scouts" started in 1945 as a radio show. From 1948 to 1958 the show was transferred to the next big thing, TV, but continued to be simulcast on the radio. From 1949 to 1959, he also broadcast "Arthur Godfrey and his Friends," also known as simply the "Arthur Godfrey Show," which featured many of the "Talent Scout" winners including Pat Boone, Eddie Fisher, Tony Bennett, Connie Francis, Patsy Cline—and Julius LaRosa. LaRosa, in fact, triggered a downturn in Arthur Godfrey's popularity in 1953 when Godfrey fired him on-air, without warning, for exhibiting too much independence or, as Godfrey put it, "a lack of humility." That was not how underlings were supposed to be treated in those days, and it didn't sit well with Godfrey's fans. But during his radio days and his early years on TV, everything he touched turned to gold. He raked in millions of dollars for his bosses and as a result was probably the first media star to become truly wealthy.

It is sometimes said that apart from being an uber-genial host, Arthur Godfrey was devoid of any special talent. This isn't strictly true; he was an accomplished ukulele player who even gave on-air ukulele lessons. Too bad color TV had not yet come in; black-and-white could never do full justice to Godfrey's shock of red hair and his colorful Hawaiian shirts as he strummed his big baritone ukulele. Through him, the ukulele found a welcome in millions of American homes.

The second contributor to the postwar ukulele craze was an Italian luthier named Mario Maccaferri, until then best known for designing the guitar played by the legendary Romani musician Django Reinhardt. Maccaferri was also a classical guitar virtuoso in his own right, perhaps in a league with Segovia, until a freak swimming accident in 1933 injured his right hand and ended his performing career forever. He continued making stringed instruments, however, and also started a business in France making high-quality reeds for woodwind instruments.

In 1939, Maccaferri fled with his family from war-torn Europe to the U.S., struggling to maintain his reed-making business (now christened the French American Reed Manufacturing Company) in the Bronx, despite a wartime shortage of the French cane he needed to make his product. Then a visit to the NY World Fair introduced him to a new concept—plastics!— and fired his imagination. He started experimenting with polystyrene reeds and found that indeed, they worked pretty well, especially when wooden reeds were unavailable. Endorsements from clarinetist Benny Goodman and other Big Band musicians brought him a good customer base. Emboldened by his success, he went on to found his own molding and manufacturing company, Mastro Industries, which produced cheap plastic versions of everything from clothespins to toilet seats.

Everything went well until it didn't. Eventually, a combination of increased competition, increased demand for more upscale fixtures, and increased availability of alternative construction materials started making serious inroads, and Mastro Industries found itself on the skids.

That's where things stood when a chance poolside encounter between Mario Maccaferri and Arthur Godfrey in a Florida hotel rocked both their worlds—and ended up putting ukuleles into the hands of millions of American kids including mine. The two men knocked back a few drinks, played a couple of impromptu duets, and bemoaned the lack of affordable mass-produce-able ukuleles. Maccaferri had long dreamed of making a plastic uku-lele but lacked the capital to proceed without some promise of success. When Godfrey replied that he could sell a million of them, Maccaferri was inspired to revive his dream and redouble his efforts to finance this new venture.

Although Maccaferri's earlier attempts to make a plastic gui-tar and his subsequent attempts to make a plastic violin were neither commercial nor artistic successes, it turned out that plastic was well suited to producing an instrument-quality uku-lele. After extensive research, he settled on Dow Styron, which gave him the warm wood-like tone he was seeking, and strung it with nylon strings made by DuPont. He called his creation the Islander and packaged it with a pick, a tuning tool, instructions, and a songbook. It sold for $5.95.

Toy plastic ukuleles like Mattel's Uke-a-Doodle had done well, but the Islander was no toy. Arthur Godfrey, who as noted would never embrace a product he wasn't willing to use himself, tried it and fell in love. He began promoting it on his shows and the orders poured in. In the end, more than nine million ukuleles were manufactured and sold. Godfrey never asked Maccaferri for a penny; he could well afford to indulge his own dream of making the ukulele a household item. A broad grin would undoubtedly have lit up the Old Redhead's freckled face to see me and a dozen other student ukulele players in an onstage strum-fest during a school assembly—a scene that was probably repeated hundreds of times in school auditoriums across the na-tion during the heyday of the ukulele.

❉ ❉ ❉

Look closely at my photo and you'll see an odd gadget on the neck of my ukulele. At first I thought it might be a capo—basic-ally a clamp used to shorten the strings evenly, enabling you to play in different keys using the same fingering—so you only have to learn one set of chords. (Call it a "cheater" at your peril!)

But no, it was something that potentially simplified the process even further—the Chord Master, an automatic chording device invented and patented by Maccaferri that allowed you to play six basic chords with the push of a button. For an extra dollar you could get a Chord Master for your Islander, as obviously my parents did.

It all comes back! The Chord Master was supposed to be attached with two rubber bands, and I sort of recall having trouble making the darn thing stay put—probably more a commentary on my technique than on the device itself. Again stirring up some long-dormant memories, I think there was an additional reason for my Chord Master woes: The buttons are labeled D7, B7, and G on the upper row and D, A7, and E7 on the lower row—indicating that the Islander was supposed to be tuned not to G-C-E-A, as I did, but rather a whole tone higher, to A-D-F#-B. That had to be confusing to anyone who was looking for the chords you'd expect to use in the key of C.

For whatever reason, in the end I ditched the Chord Master. Instead, I learned a handful of chords and then limited my repertoire to songs that required only those chords.

* * *

Predictably, imitators started cutting into Mastro's market, and in response additional models, both larger ukes and the smaller "ukette," proliferated in the wake of the Islander's success. And then, with the advent of rock music, the popularity of the ukulele took a nosedive. Arthur Godfrey had once reassured parents, "If a kid has a uke in hand, he's not going to get into much trouble." Where's the fun in that? Kids eager to cultivate a "bad boy" image followed their musical heroes and took up the guitar, not the ukulele. If you were inclined to write the ukulele's obituary, you could do worse than to choose February 9, 1964 as the day it officially died, with the Beatles' American debut on the Ed Sullivan Show (though in fact the Beatles loved the ukulele and occasionally played it onstage). Herbert "Tiny Tim" Khaury probably put the final nail in its coffin in 1968 with his popular but wildly campy falsetto rendition of "Tiptoe through the Tulips," stripping the poor old uke of whatever shred of dignity it might once have had and masking its genuine musical versatility.

But the ukulele just won't stay dead. It has enjoyed at least two revivals since then. The first was in the 1990s, when a new generation of instrument makers started appealing to a new generation of musicians, most of whom had forgotten Tiny Tim and maybe even the Beatles. The second is now in progress, fueled by the rise of YouTube ukulele artists like Hawaii native Jake Shimabukuro, whose videos routinely go viral. And with the latest latest additions to the ukulele family, the electric ukulele and the acoustic electric ukulele that sounds great either plugged or unplugged, the uke is now busily burrowing its way into the rock scene.

My once-cherished Islander has long since disappeared into the mists of time, and until recently, I hadn't touched a uke since the 1950s. E-Bay to the rescue; I was able to purchase an exact duplicate, and having been cleaned up, repaired, and restrung, it now awaits my efforts to get even close to my former modest level of virtuosity. Who says you can't go home again?

ACKNOWLEDGEMENTS

I begin this section with gratitude and enthusiasm for all the help I've received along the way, not only during the actual writing of this book but also during the preceding five years while I was writing the blog that served as the basis for much of its contents. I also approach it with trepidation about un-deserved omissions, either through lapse of memory or because I didn't recognize the importance of someone's contribution at the time it was given (remembering, again, that this book, unbe-knownst to me, has been five years in the making). Expect this section to be longer in future editions!

To start at the beginning, sometime in the late 1990s my high school friend Mary Joyce Carlson read a notice somewhere —in her late husband's Massachusetts Institute of Technology alumni magazine, she believes—about Robert Buderi's (then) new book, entitled *The Invention that Changed the World: How a Small Group of Radar Pioneers Won the Second World War and Launched a Technological Revolution*. Because she knew my father had been a pioneer in the development of radar, she thought I might be interested in the book. I was! Here was my chance to learn more about Project Diana.

When I got my hands on it, however, I found that the book was not about Project Diana at all, but rather about (duh!) the Massachusetts Institute of Technology Radiation Laboratory. Project Diana took up just over three pages, and that only be-cause DeWitt and his team had stolen a march on the Rad Lab by getting there first. But even though Project Diana was peripheral to his research, Bob was most gracious on that and subsequent occasions, allowing me to quote from his book and offering general encouragement. He also did me the inestimable favor of putting me in touch with a local history buff named Fred Carl, the sparkplug behind the InfoAge Science History Learning Center and its tireless efforts to preserve the long and checkered past of the Camp Evans site. In 1993, with the end of the Cold War, the Department of Defense had decided to close many of its military bases, and Camp Evans was on the hit list. But for Fred's remarkable vision, the entire Camp Evans campus with all its rich heritage would have been plowed under. I doubt my Project Diana website, its blog, or this book could exist without direct and indirect help from Fred and his crew, including John Cervini, Ray Chase, Gloria Kudrick (Editor of the *InfoAge Marconigraph*), Al Kerecman, Lori Lauber (who ably presides over the Project Diana exhibit), Dan Lieb, Mike Ruane (CEO), and probably others.

The following individuals were generous in sharing their technical expertise:

Al Klase and Ray Chase, historians at the New Jersey Antique Radio Club's Radio Technology Museum at InfoAge, helped me immeasurably in getting the Armstrong story straight, explaining Walter McAfee's contribution to Project Diana, and other technical details.

Chris O'Connor is a retired Colonel now living in Virginia, but during his military career he worked closely with mobile radars from the program office at Camp Evans. He provided me with one of the most helpful reader suggestions I ever received on my blog, directing me to a recently declassified document that finally enabled me to resolve the mystery of why the Army had largely orphaned Project Diana—leading to my essay "The Moon Enters the Cold War."

Greg Wright, who works in Wireless Communication Research at Nokia Bell Labs in Holmdel, New Jersey, introduced me to the mysteries of moon regolith and helped me sort out the issue of what Project Diana taught us about the moon itself that might have been useful in guiding that other, even more famous moonshot, Apollo 11, as described in "A Tale of Two Moonshots."'

Early on I was fortunate enough to find my way to David Mofenson, my fellow Project Diana legacy and son of Jack Mofenson. David was about two weeks older than I and my earliest playmate, as documented in many family photos. I have an old home movie of David and me eating together when we were barely old enough to hold our spoons—only David spent more time feeding me than he did feeding himself, that's the kind of good and generous person he was even then. (I'm sorry to report I was more intent on feeding myself than on reciprocating.) David was a Boston area lawyer and getting ready to retire, eagerly anticipating a long-deferred trip to France with his wife when we first connected, but he willingly made time to interact with me, even seeking updates about the Stodola clan to share with his mother. In fact, he was so enthusiastic about my plan to launch a Project Diana website that we started to discuss sharing responsibility for the content. To my great sorrow, he died suddenly and unexpectedly before we could make any progress with this plan. Rest in peace, David.

After several years of fruitless searching, I finally connected via ancestry.com with another Project Diana legacy, Diana Webb, daughter of Harold Webb (and yes, she was indeed named after Project Diana), who helped me flesh out my description of her dad and has fed me quite a bit of other useful material from her archives. Irv Cantor, son of Gilbert Cantor, who described his role on Project Diana as "King Stodola's tool-in-hand," corroborated my recollections of waiting outside the Lab but never going in. I am now in the process of reaching out to other Project Diana legacies in the hopes of growing our little club while we're still here to share our memories.

Speaking of ancestry.com, the people who hang out there are generally the most helpful people you could ever hope to meet. I spent many years making little headway in determining my paternal grandfather's family history; Jewish genealogy is notoriously difficult to unravel due to widespread endogamy (within-group marriage), and tracing his forebears has caused me more frustration than my other three grandparents combined. I still have a lot to learn, but I will be forever grateful to Jodi Hom for much of what I have uncovered so far; she gave me exactly the breadcrumbs I needed to figure out the circumstances under which my great great grandparents emigrated to the U.S. in the mid-19th century, a story recounted in the essay titled "When Edwin Met Beatrice."

If Professor Google doesn't know the answer to your question, ask a librarian. Librarians who patiently responded to my questions include Jennifer S. Comins, Archivist of the Carnegie Collections in the Rare Book and Manuscript Library at Columbia University, where the Armstrong papers are housed, and Dave McMullin, Music Librarian at the New York Public Library for the Performing Arts, home of the Henry Holden Huss papers.

My second-cousin-once-removed Ian King, the unofficial King family genealogist, knew little about what had become of my great grandfather Arthur and his descendants but a great deal about other King relatives. During our 2019 visit to the UK he miraculously extracted, from a briefcase overflowing with documents and memorabilia, a set of color photos of the King family's Boys' Butterfly Collection, which still exists (somewhere) and which provided the context for my father's boyhood butterfly-collecting expeditions with his grandfather.

Lisa Kozenko, Associate Professor of Music at Ball State University in Muncie, Indiana, whose doctoral dissertation covered the music scene in New York during the peak of my grandfather's concert career, was helpful in checking my conclusions and steering me to additional information on Harry Holden Huss.

Melinda Ponder, biographer of Katharine Lee Bates, reality-checked my father's assertion that Bates was an elocution student of his mother.

My cousin-in-law Alan Wald, Emeritus Professor of English Literature and American Culture at the University of Michigan, provided information I'm pretty sure I'd never have otherwise unearthed on the life of Alphonse Tonietti, who appears in my essay on Shark River Hills, and was also very helpful on my father's brush with literary history.

Sandra Chaff, who presided over the Archives of the Medical College of Pennsylvania when I was working on the Oral History Project on Women Physicians and who has been my friend and Scrabble partner ever since, was able to dig up information on Dr. Helen Jones that had eluded me on Google.

Linda Howe Steiger, my close friend since graduate school and also a novelist and teacher of memoir-writing, was incautious enough to venture a comment on an early version of this book and discovered that she had somehow backed into serving as both beta reader and editor, for which I am profoundly grateful—even though I'm certain there's nothing I could ever do to make her love my essay on ham radio. ("What the heck is CQ?!")

Anne Curzan, Dean of Literature, Science, and the Arts at the University of Michigan and my consultant on usage, endorsed "moon" not "Moon" and gave me permission to use "the life and times of" in my subtitle even though Project Diana is not a living being so doesn't technically have a life. She's heard weekly on NPR's feature "That's what they say," so she should know.

I like to think I remember my childhood very well, but my Shark River Hills BFF Joyce Greco is amazing and provided many additional details when I thought I'd already covered it all.

A shout-out to Bill Evers and Helen Evers, my friend Sally's younger siblings, who supplied me with wonderful family photos as well as many memory jogs.

Two Facebook Shark River Hills nostalgia sites, "You Know You're from Shark River Hills, NJ If..." and "Shark River Hills Reunion," brought forth surprising insights along the way—looking at you, Randy Rossman, whom I've never met but who obviously shares my love of old newspapers and SRH trivia. The Facebook Earth-Moon-Earth (EME) Radio Communications group also includes a cadre of history buffs who regularly cheer me on.

The current owner of my childhood home on Pinewood Drive was kind enough to let us wander at will through the house, confirming some of my memories and confuting others. If his wife had been home at the time I'm not sure we'd have been so lucky —so I won't blow his cover by revealing his name!

First, last, and in between, beyond any possibility of repayment, I am indebted to my family:

My sisters Leslie Darland and Sherry Rapport were invaluable memory-extenders, both about our dad and about Shark River Hills. My brother Bob was only three when we left New Jersey, but he was very helpful to me in reconstructing my father's life, particularly the years after my mother's death, during which he spent a lot more time with our dad than any of the rest of us did. I have been buoyed throughout by the unflagging enthusiasm of all my siblings for my efforts to honor Project Diana and our father.

My daughters Julie Pomerleau and Aimee Pomerleau, and my grandchildren Augie Pomerleau and Claudia Stafford, were of course not around when the events recounted in this book took place. For that very reason, I regard them as the primary inspiration for this effort and as the readers to whom I most wish to pass along this information. If no one else ever reads this book, I'd still have done it for them. These are their stories, too.

My husband and life companion Ovide Pomerleau, who is very good at coaxing people into telling him their stories, had many long conversations with my dad over the course of innumerable family gatherings and was able to provide not only technical details on my father's accomplishments but also insights into his professional career and opinions about life in general that I missed either because I lacked the bandwidth to grasp them at those busy moments or was otherwise occupied basting the turkey or frosting the cake. He also provided enough technical support based on his experience as an amateur radio operator to keep me out of trouble on topics relating to radar and radio; I ignored his cautions seldom, but at my peril! Finally, he willingly read every essay at least once and often two or three times after major rewrites, which in fact was quite typical; he also served as a beta reader for the book as a whole. For all that and more, I love him to the moon and back.

ABOUT THE AUTHOR

Cindy Stodola Pomerleau

Cindy Stodola Pomerleau was just shy of three years old when the U.S. Army successfully bounced radio waves off the moon—the first-ever extraterrestrial communication, the birth of radar astronomy, and the opening salvo in the Cold War. She was born on the Jersey coast for the same reason as Project Diana—her father, as Scientific Director of the Project, was intimately involved in both events. Like Project Diana, she was named for the goddess of the moon (in her case Cynthia, a nickname for Artemis, the Greek version of Diana, who was born on Mt. Cynthos). Project Diana is baked into her DNA.

She holds a Ph.D. in English literature from the University of Pennsylvania and a master's degree in psychology from the University of Hartford. Her professional background includes directing the Oral History Project on Women in Medicine at the Medical College of Pennsylvania and later serving as Research Professor in the University of Michigan Department of Psychiatry, where she studied smoking and nicotine dependence (1985-2009). Now retired, she lives in Ann Arbor, Michigan with her husband of 55 years, Ovide Pomerleau. To learn more about her books, articles, and scientific writings, visit her website at www.cindypomerleau.com.

www.ingramcontent.com/pod-product-compliance
Lightning Source LLC
Chambersburg PA
CBHW070325220526
45467CB00001B/31